SERIES 1—NINTH EDITION

TECHNICAL DRAWING PROBLEMS

FREDERICK E. GIESECKE
Late Professor Emeritus of Drawing
Texas A & M University

ALVA MITCHELL
Late Professor Emeritus of Engineering Drawing
Texas A & M University

HENRY CECIL SPENCER
Late Professor Emeritus of Technical Drawing;
Formerly Director of Department
Illinois Institute of Technology

IVAN LEROY HILL
Professor Emeritus of Engineering Graphics;
Formerly Chairman of Department
Illinois Institute of Technology

JOHN THOMAS DYGDON
Professor of Engineering Graphics,
Chairman of the Department,
and Director of the Division of Academic Services
and Office of Educational Services
Illinois Institute of Technology

JAMES E. NOVAK
Associate Director/Executive Officer,
Office of Educational Services
Illinois Institute of Technology

Macmillan Publishing Company
NEW YORK
Maxwell Macmillan Canada
TORONTO
Maxwell Macmillan International
NEW YORK OXFORD SINGAPORE SYDNEY

Laser typesetting by **Ewing Systems**, 409 W. 24th St., Suite #14, New York, NY 10011. Printed in the United States of America.

Earlier edition entitled *Technical Drawing Practice*, by Frederick E. Giesecke, Alva Mitchell, and Henry Cecil Spencer, copyright 1934 by the Technical Book Company. Earlier editions entitled *Technical Drawing Problems*, by Frederick E. Giesecke, Alva Mitchell, and Henry Cecil Spencer, copyright 1936 and 1947 by Macmillan Publishing Co., Inc. Earlier editions entitled *Technical Drawing Problems*, Series 1, by Frederick E. Giesecke, Alva Mitchell, Henry Cecil Spencer, and (1967 and 1976 editions) Ivan Leroy Hill, © 1959 and copyright © 1967, 1976, 1981 by Macmillan Publishing Company. Copyright renewed 1962 and 1964 by Hulda Gruene Giesecke, R. Howard Mitchell, and Henry C. Spencer.

Macmillan Publishing Company
866 Third Avenue, New York, New York 10022

Macmillan Publishing Company is part of the Maxwell Communication Group of Companies.

Maxwell Macmillan Canada, Inc.
1200 Eglinton Avenue East
Suite 200
Don Mills, Ontario M3C 3N1

ISBN: 0-02-342770-1

Printing: 1 2 3 4 5 6 7 8 Year: 1 2 3 4 5 6 7 8 9 0

Preface

Technical Drawing Problems, Series 1, may be used with any good text on technical drawing or engineering graphics, but it has been designed especially to be used with the text *Technical Drawing* (Ninth Edition, Macmillan, 1991). References in the instructions are to this text.

In the Instructions detailed directions are given for each sheet, together with references to the text. *It is expected that the student will read these instructions and study the references before each sheet is started.*

The problems included here are designed to cover the important basic principles of technical drawing, but are not intended to constitute a complete course in themselves. It is expected that the instructor will supplement these sheets, particularly in the later phases of the work, with problems assigned from the text. In general, the types of work covered here are those deemed most suitable for this kind of presentation, while those to be covered as supplementary problems are best adapted to work on larger sheets in the manner of commercial practice. A number of problems are presented on vellum to provide the student with experience more closely related to industrial practice. Each problem is now in accord with the new Ninth Edition of *Technical Drawing*.

A feature of this revision is the minimum use of fractional inch dimensions and the emphasis on metric and decimal-inch dimensions now used extensively in industry. A decimal and millimeter equivalents table and appropriate full-size and half-size scales are provided inside the front and back covers for the student's convenience.

Emphasis on freehand sketching continues to be an important feature of this workbook, and many problems are designed specifically to be drawn freehand. The instructor will also find a number of the instrumental drawing problems equally suitable for freehand sketching.

All of the problems are based upon actual industrial designs, and their presentations are in accord with the latest ANSI Y14 American National Standard Drafting Manual and other relevant ANSI standards. Moreover, a special effort has been made to present problems that are thought provoking rather than requiring a great deal of routine drafting.

Byron L Johnson
851-3523

In response to the increased usage of computer technology for drafting and design, a number of problem sheets in computer-aided drafting (CAD) have been included. The new problem sheets on detail drawings are presented to provide practice in making regular working drawings of the type used in industry. These are suggested for solution either by a computer-aided drafting system or by traditional drafting methods.

The authors wish to express appreciation to their colleagues for many valuable suggestions and to numerous industrial firms who have so generously cooperated in supplying problem material. Comments and criticisms from users of this workbook will be most welcome.

Ivan Leroy Hill
Clearwater, FL

John Thomas Dygdon
Illinois Institute of Technology
Chicago, IL

James E. Novak
Illinois Institute of Technology
Chicago, IL

Contents

Worksheets

Vellums

Printed on vellum, these numbered sheets appear at the back of this workbook, after Sheet 109.

Instructions

References to Ninth Edition of *Technical Drawing (1991)*
by Giesecke, Mitchell, Spencer, Hill, Dygdon, and Novak.

Throughout this workbook alternative dimensions, often not the *exact* equivalents, are given in millimeters and inches. Although it is understood that 25.4 mm = 1.00", it is more practical to use approximate equivalents such as 25 mm for 1.00"; 12.5 or 12 mm for .50"; 6 mm for .25; 3 mm for .12"; etc. Exact equivalents should be used when accurate fit or critical strength is involved.

In general it will be found that the following leads are suitable for instrumental drawing: a 4H for construction lines and guide lines for lettering; a 2H for center lines, section lines, dimension lines, and extension lines; and an F for general linework and lettering. All construction lines on problems should be made *lightly* and *should not be erased.*

Sheet 1. Vertical Capital Letters. References: §§4.1–4.18, 4.24. In the upper half of this sheet, fill in the lines with neat freehand copies of the letters shown, using an HB lead, and continuing approximately the spacing indicated. The vertical guide lines will help you to keep your letters vertical; they are not to be used to space the letters. These large letters may be sketched lightly first, and corrected where necessary, before being made heavy with the strokes shown. Do not show the numbers and arrows on your letters, and do not draw any additional grids or guide lines.

In the lower half of the sheet under each line of lettering, make a copy as nearly like the letters shown as you can, using the guide lines provided. Use the strokes as shown, forming each character with single strokes (no preliminary sketching).

Sheet 2. Vertical Capital Letters. References: §§4.1–4.18, 4.24. Letter this sheet in the manner described for Sheet 1. This will complete the alphabet, and you should know all the strokes and proportions of the letters. If you do not, practice the letters on cross-section paper, making each letter six squares high until you know them thoroughly. Do not show the numbers and arrows on your letters, and do not draw any additional grids.

Sheet 3. Vertical Numbers. References: §§4.16–4.18, 4.20. In the upper portion of the sheet, fill in the lines with neat freehand copies of each numeral, using a sharp HB lead and continuing the spacing approximately as indicated. Unless otherwise assigned, make each numeral four times. These large numerals may be sketched lightly first, and corrected where necessary, before being made heavy with the strokes shown. Notice that all numerals except the 1 are five units wide; the zero is therefore one unit narrower that the letter O.

To the right of each of the large fractions and decimals fill in the lines with two copies of the examples, spaced about like the larger characters above. Form each character with single strokes (no preliminary sketching) as indicated on the large numerals.

In the lower portion of the sheet, below each line of fractions or whole numbers, letter two lines of the same characters. The vertical guide lines will help you to keep your letters vertical; they are not to be used to space the letters. Be very careful, particularly on the smallest fractions, not to allow the numerals to touch the fraction bars, Fig. 4.29 (a). Use a sharp HB lead for the larger numerals, a sharp F lead for the smaller numerals.

Sheet 4. Vertical Capital Letters. References: §§4.1–4.18, 4.24. Draw extremely light vertical guide lines at random over the entire sheet (draw lines the full length of the sheet). Under the given line, letter the words as nearly like the originals as you can, using a sharp HB lead for the larger letters and a sharp F lead for the smaller letters. Accent the ends of all strokes to give definiteness, Fig. 4.17. Pay particular attention to spacing letters by the eye so that the background areas will appear approximately equal and the words are spaced far enough apart to insert a letter O between them, §4.24.

Sheet 5. Vertical Lowercase Letters. References: §§4.14, 4.16, 4.17, 4.21, 4.22, 4.24. Except where grids or boxes are already shown, draw light vertical guide lines at random over the entire sheet. At the right of each of the large letters, fill in the lines with neat copies, spaced by eye approximately as indicated. Use a sharp HB lead for the largest letters and a sharp F lead for the two smaller sizes, and make each letter directly with a series of single strokes, as shown in the given letters.

Sheet 6. Inclined Capital Letters. References: §§4.1–4.17, 4.19, 4.24. In the upper half of the sheet, fill in the lines with neat freehand copies of the letters shown, using an HB lead, and continuing approximately the spacing indicated. The inclined guide lines are not to be used for spacing letters, but as references to help you keep your letters inclined at a uniformed angle. These large letters may be sketched lightly first, and corrected where necessary, before being made heavy with the strokes shown. Do not show the numbers and arrows on your letters and do not draw any additional grids or guide lines.

In the lower half of the sheet, under each line of lettering, make a copy as nearly like the letters shown as you can, using the guide lines provided. Use the strokes shown, forming each character with single strokes (no preliminary sketching). Space words far enough apart to insert a letter O between them, §4.24.

In making inclined letters, it is very important to have a uniform inclination on all characters. This may be difficult on letters A, W, V, X, and Y. For each of these, balance the letters symmetrically about an imaginary inclined center line. This will make the right side of the A and the left side of the V nearly vertical, as well as strokes 1 and 3 of the W.

Sheet 7. Inclined Capital Letters. References: §§4.1–4.17, 4.19, 4.24. Letter this sheet in the manner described for Sheet 6. This will complete the alphabet, and you should know all the strokes and proportions of the letters. If you do not, practice the letters on scratch paper, making them from memory and then comparing with the strokes and proportions given in your text. Do not show the numbers and arrows on your letters, and do not draw any additional grids.

Sheet 8. Inclined Numerals. References: §§4.16, 4.17, 4.19, 4.20. Draw extremely light inclined guide lines at random over the entire sheet, except where grids are shown. In the upper portion of the sheet, fill in the lines with neat freehand copies of each numeral, using a sharp HB lead and continuing the spacing approximately as indicated. The large numerals may be sketched lightly, and corrected where necessary, before being made heavy with the strokes shown. Notice that all numerals except the 1 are five units wide; the zero is therefore one unit narrower than the letter O.

To the right of each of the large fractions and decimals, fill in the lines with copies of the examples, spaced about like the larger characters above. Form each character with single strokes (no preliminary sketching), as indicated on the large numerals.

In the lower portion of the sheet, below each line of fractions or whole numbers, letter two lines of the same characters. Be very careful, particularly on the smallest fractions, not to allow the numerals to touch the fraction bars, Fig. 4.29 (a). Use a sharp HB lead for the larger numerals and a sharp F lead for the smaller numerals.

Sheet 9. Inclined Capital Letters. References §§4.1–4.17, 4.19, 4.24. Draw extremely light inclined guide lines at random over the entire sheet. Under each given line, letter the words as nearly like the originals as you can, using a sharp HB lead for the larger letters and a sharp F lead for the smaller letters. Accent the ends of all strokes to give definiteness, Fig. 4.17. Pay particular attention to spacing letters by eye so that the background areas are approximately equal and the words are spaced far enough apart to insert a letter O between them, §4.24.

Sheet 10. Inclined Lowercase Letters. References: §§4.14, 4.16, 4.17, 4.21, 4.23, 4.24. Except where grids or boxes are already shown, draw light inclined guide lines at random over the entire sheet. At the right of each of the larger letters, fill in the lines with neat copies, spaced by eye approximately as indicated. Use a sharp HB lead for the largest letters and sharp F lead for the two smaller sizes. Make each letter directly with a series of single strokes, as shown in the given letters.

Sheet 11. Conventional Lines. References: §§2.1–2.16, 2.24–2.43, 2.46, 2.52, 4.18, 13.7.

Space 1. Using the T-square, complete the conventional lines to fill the right half of the space and to match the lines in the left half. Use a sharp F lead with a conical point, Fig. 2.8, for the visible line, hidden line, and cutting-plane line; a 2H lead for the center line, dimension line, extension line, section line, and phantom line. Make all lines **black** but of the correct widths.

Space 2. Draw the view of the Anchor Slide full size, locating the view by the starting center point A, as indicated. Your final lines should approximate those at the left in Space 1, with three distinct thicknesses of lines. See inside of back cover of the workbook for metric and decimal-inch scales. If assigned, use convenient approximate decimal-inch equivalent dimensions for the layout.

Sheet 12. Use of T-square and Triangles. References: §§2.5–2.17, 2.20–2.28, 2.32, 2.46. For *Spaces* 1–4, locate the center of the space by drawing light diagonals (draw only light dashes crossing at the center). Through this point draw a construction line perpendicular (90°) to the required lines. Along this line set off the desired spaces from the center point. Draw lines to fill the spaces as follows.

Space 1. Horizontal visible lines 12.5 mm (.50") apart. Use scale in position shown in Fig. 2.54, Step III, and draw lines shown in Fig. 2.21.

Space 2. Vertical hidden lines, 12.5 mm (.50") apart. Use scale in position shown in Fig. 2.54, Step II, and draw lines as shown in Fig. 2.22.

Space 3. Inclined section lines 30° with horizontal, sloping upward to the right, and spaced 12.5 mm (.50") apart. Along the construction line drawn through the center point and perpendicular to the required lines, set off the desired spaces from the center point. (Some division points will fall outside the rectangle.) Draw the required lines through these points and at right angles to the construction line. See Fig. 2.23, E and L.

Space 4. Inclined center lines 60° with horizontal, sloping downward to the right, and spaced 12.5 mm (.50") apart. Locate the center of the space as before, draw a construction line through the center point and perpendicular to the required lines. Set off distances on this line, making one mark cross through the center; then draw the required lines at right angles to this line through these points. See Fig. 2.23, E and L.

Space 5. Draw cutting-plane lines parallel to given line and at 12.5 mm (.50") intervals to fill the space.

Space 6. Draw visible lines 12.5 mm (.50") apart and perpendicular to the given line. Arrange so that one visible line passes through center of space.

Sheet 13. Technical Sketching. References: §§6.1–6.10, 6.24–6.26. Using an HB lead, sketch the given views on the enlarged grids as indicated, starting at points A. Make the visible lines the proper width and **black** to make the views stand out from the grid. Add center lines as shown.

Sheet 14. Technical Sketching. References: §§6.1–6.10, 6.24–6.26. Using an HB lead, sketch the given views on the enlarged grids as indicated, starting at points A. Make all final lines the correct widths and **black**.

Sheet 15. Mechanical Drawing. References: §§2.2–2.19, 2.24–2.28, 2.31, 2.32, 2.39–2.43, 2.46, 2.47, 7.4.

Space 1. Draw sharp **black** lines, the weight as section lines, to the lengths and scales indicated, using the architects, engineers, or metric scale. Draw each line from the marks shown, measuring from the middle of the vertical starting line. At the end of each line, draw a *sharp* accurate vertical dash to indicate the *exact length. Your measurements must be accurate.* At K through M determine the scales and lengths of lines, and record the scales and lengths in the spaces provided. Make lettering the same height as given for A through J.

Space 2. Using the starting points A and B, lay out the views lightly with construction lines, and obtain instructor's approval. Omit dimensions. Darken the lines, using an F lead for visible and hidden lines and a 2H lead for center lines.

Sheet 16. Mechanical Drawing. References and instructions same as for Sheet 15.

Sheet 17. Geometry of Technical Drawing. References: §§2.32–2.43, 5.1–5.8. In the following constructions, draw light *accurate* construction lines, using a sharp 4H lead in both pencil and compass. Darken the required lines with a sharp F lead. For example, in Space 1 the compass arcs should be extremely light (hard lead), while the *required* bisector should be sharp but **black** (F lead). If assigned, letter all constructions as shown in the corresponding figures in the text.

Space 1. Reference: Fig. 5.8.
Space 2. Reference: Fig. 5.8.
Space 3. References: §2.39, 2.40, Figs. 5.14, 5.15.
Space 4. References: Figs. 5.10, 5.11.
Space 5. Reference: Fig. 5.18 (b). Check by (e).
Space 6. Reference: Fig. 5.32 (a).

Sheet 18. Geometry of Technical Drawing. General references and instructions same as for Sheet 17.

Space 1. Reference: Fig. 5.22 (a). Check by (b).
Space 2. Reference: Fig. 5.23 (a).
Space 3. Reference: Fig. 5.23 (c).
Space 4. Reference: Fig. 5.26 (b). Check by Fig. 5.25 (a).
Space 5. Reference: Fig. 5.26 (c) or (d).
Space 6. Reference: Fig. 5.28 (b).

Sheet 19. Geometry of Technical Drawing—Tangencies. General references and instructions same as for Sheet 17.

Space 1. Reference: Fig. 5.34. Construct point of tangency for lower line.
Space 2. Reference: Fig. 5.36 (b). See also §5.33.
Space 3. Reference: Fig. 5.38 (a).
Space 4. Reference: Fig. 5.38 (b).
Space 5. Reference: Fig. 5.37 (a).
Space 6. Reference: Fig. 5.39 (a).

Sheet 20. Geometry of Technical Drawing—Tangencies. General references and instructions same as for Sheet 17.

Space 1. Reference: Fig. 5.39 (b).
Space 2. Reference: Fig. 5.40 (a).
Space 3. Reference: Fig. 5.40 (b).
Space 4. Reference: Fig. 5.36 (a). Show construction for point of tangency.
Space 5. Reference: Fig. 5.41 (a).
Space 6. Reference: Fig. 5.41 (b).

Sheet 21. Geometry of Technical Drawing—Ellipses. General references and instructions same as for Sheet 17.

Space 1. Reference: Fig. 5.56.
Space 2. Reference: Fig. 5.64 (c).
Space 3. References: Figs. 5.47–5.50

Sheet 22. Geometry of Technical Drawing—Parabola and Hyperbola. General references and instructions same as for Sheet 17.

Space 1. Reference: Fig. 5.57 (b).
Space 2. Reference: Fig. 5.57 (c).
Space 3. Reference: Fig. 5.60 (b).
Space 4. Reference: Fig. 5.61 (a) or (b).

Sheet 23. Technical Sketching. References: §§1.1–1.10, 6.1–6.31. Using an HB lead, sketch the views or isometrics as indicated. Make lines *clean-cut* and **black.** Omit hidden lines from the isometrics. Take great care to sketch isometric ellipses correctly, §6.13.

Sheet 24. Technical Sketching. References and instructions same as for Sheet 23. Observe the rule: *Parallel lines on the object will appear parallel both on the views and in the isometric.*

Sheet 25. Sketching Missing Lines. References: §§6.1–6.7, 6.18–6.31, 7.12–7.18. In each of the twenty-four problems on this sheet, one or more lines have been omitted intentionally. In Prob. 6, two views are incomplete; in each of the others only one view is incomplete. Supply the missing lines freehand.

Make your lines the same width and blackness as the printed lines, and make hidden dashes the same length as the given dashes. Be careful to make correct junctures of hidden-line dashes, Fig. 6.43.

If you have difficulty in solving any problem, make a small isometric sketch as shown accompanying the first four problems, or make a small model of soap or clay, §7.14.

Some problems have more than one solution. In such cases, any solution that is correct will be accepted.

Sheet 26. Sketching Missing Lines. References and instructions same as for Sheet 25. In Probs. 3 and 11, two views are incomplete; in each of the others only one view is incomplete. If you feel that you need further practice in multiview drawing, sketch on cross-section paper problems selected from Figs. 6.51–6.54.

Sheet 27. Multiview Projection—Optional Views. References: §§6.5–6.8, 6.18–6.31, 7.12–7.18. In each of the three spaces on this sheet is given the top or front view of a series of simple objects. Design the missing views so that no two objects will be alike. Make the drawings freehand. Add all necessary center lines. No visible line must be imagined as the edge view of a thin piece of sheet metal or other material. Create and draw reasonable shapes; avoid impractical forms.

Sheet 28. Surface and Edge Identification. References: §§6.12, 6.18, 6.30, 6.31, 7.12–7.28

Space 1. The various surfaces are identified by letters on the isometric drawing and by numbers on the three views. In the table, letter with a sharp F lead the numbers corresponding to the letters given at the left, using the guide lines shown.

Space 2 and 3. In the tables, letter with a sharp F lead the numbers of the corresponding surfaces and edges, using the guide lines shown. Give the numbers taken from a view as read in a clockwise direction, starting at the upper left in each series. For example, in the top view of Prob. 2, the triangular surface is **3–4–7**, not **4–7–3** or **7–4–3**.

Sheet 29. Technical Sketching. References: §§6.4–6.10, 6.18–6.31. Using an HB lead, sketch the three required views as indicated. Position the front view at the corner mark A. The isometric is drawn half scale. In sketching the three views, follow the procedures explained in §6.28. First, block in the views lightly freehand and obtain the instructor's approval before darkening the lines. No mechanical aids are to be used, except for guide lines in the title strip. Your lines must be *clean-cut* and **black** and exhibit good freehand technique, §6.5.

Sheet 30. Missing Views and Lines. References: §§6.25, 6.26, 6.30, 6.31, 7.1–7.7, 7.12–7.33. In each problem two views are complete, and you are to draw freehand the missing view or lines. First, sketch the missing views or lines lightly with construction lines; then, when you are sure your solution is correct, darken the lines with an HB or F lead, matching your lines in both width and blackness with the printed lines. In Prob. 1 fill in answers to the questions, and in Prob. 2 add the corresponding numbers in the top view .06" or 1.6 mm high. Study §7.6 and Fig. 7.8.

If you have difficulty visualizing any problem, try sketching an isometric of it, §§6.12–6.14. Pay particular attention to the correct junctures of hidden-line dashes, Fig. 6.43.

Sheet 31. Missing Views and Lines. References and instructions same as Sheet 30. For Probs. 1, 2, and 3, study §5.34 and Fig. 7.37, and show constructions for the points of tangency. Draw all necessary center lines.

Sheet 32. Missing Views and Lines. References: §§7.12–7.28, 7.33.

Space 1. Draw the missing right-side view, using instruments.

Space 2. Lines are missing in all three views. Study the construction carefully in the pictorial drawing, and supply the missing lines using instruments. In the right-side view draw a line from A to B, and extend it to the point that corresponds to X in the pictorial drawing. In the top view locate point X, join it to point C, extend to point Y as shown in the pictorial drawing, and continue in this manner. Apply the rule given in §7.25.

Sheet 33. Missing Views. References: §§7.27–7.31, 7.33.

Space 1. Draw freehand the top view as indicated.

Space 2. Draw with instruments the front view as indicated. Plot a sufficient number of points to define the curve accurately. Sketch the curve lightly through the points with a 4H lead; then draw the curve with the F lead and with the aid of the irregular curve, §2.54.

Sheet 34. Missing Views. References: §§7.17, 7.18, 7.21–7.25, 7.33, 12.20. Supply the missing views as indicated, using instruments.

Sheet 35. Missing Views. References same as for Sheet 34. Supply the missing views mechanically as indicated. In Prob. 2 add missing lines in right-side view and front view for surface A that passes through points 1, 2, and 3. Use the principles discussed in §7.25.

Sheet 36. Optional Views and Described Shape. References: §§6.5–6.10, 6.18–6.31, 7.12–7.18. Use instruments to make full-size drawings of the required views. First, sketch the given views on cross-section paper and add the views that you propose to draw on the sheet. If necessary, draw a small isometric sketch to confirm your solution. Obtain your instructor's approval of the sketches; then draw the views on the sheet.

Space 1. Design front and right-side views that will agree with the given top view. Height of block is 25 mm (1.00"). Space views 10 mm (.40") apart. Show all center lines.

Space 2. Design top and right-side views that will agree with the given front view. Depth of block is 38 mm (1.50"). Space views 10 mm (.40") apart. Show all center lines.

Space 3. Design front and top views that will agree with given right-side view. Width of block is 60 mm (2.36"). Space views 10 mm (.40") apart. Show all center lines.

Space 4. Draw front, top, and right-side views of rectangular block 57 mm (2.24") wide, 28.5 mm (1.12"), and 35 mm (1.38") deep, spacing views 10 mm (.40") apart. Remove lower right quarter of block as seen in the front view.

As seen in the top view, the four outside corners are rounded to 10 mm (.38") R. On the two centers at the left end, two 6 mm (.24") DIA holes are drilled vertically through the block, and then counterbored to 12.5 mm (.50") DIA and to a depth of 6 mm (.24").

A slot 12.5 mm (.50") wide and 31.5 mm (1.24") overall length (as seen in the top view) is cut centrally from the right end with its length parallel to the long dimension of the block. The inner end is rounded to 6 mm (.24") R, and the true shape is shown in the top view.

Complete all views, showing all hidden lines and center lines.

Sheet 37. Missing Views. References: §7.7, 7.21, 7.27–7.29, 7.32–7.36, 12.1–12.5, 12.20, 13.17. With the right-side view given, draw the front and top views with instruments. Necessary supplementary dimensions are given in the pictorial drawing. Show all finish marks on the three views. The holes are understood to be finished; omit finish marks on them.

Sheet 38. Missing View. References: §§7.7, 7.21, 7.27–7.30, 7.33, 12.19–12.20, 15.7–15.10. Complete the top view as indicated, using instruments. The top view of the hole shown by a solid circle and a hidden circle is that of a tapped (threaded) hole, as shown in Fig. 15.8 (a). Locate the top view according to the top view of this hole. The object is understood to be finished all over, and no finish marks are necessary.

Sheet 39. Missing View. References: §§7.7, 7.17, 7.21, 7.28–7.36, 12.1–12.5, 12.20, 13.17. Complete the right-side view as indicated, using instruments. The material is cast iron. The machining is shown by the finish marks in the given views; add finish marks in the required view. Omit finish marks from the holes; they are understood to be finished. Fillets and rounds are 1.5 mm (.06") R.

Sheet 40. Missing View. References: §§7.7, 7.32–7.36, 12.1–12.5, 12.20, The top and right-side views are given; add the front view, using instruments. Add all necessary finish marks. Omit finish marks from the holes; they are understood to be finished. Include all hidden lines and center lines.

Sheet 41. Sectional Views. References: §§9.1–9.9, 9.11–9.13. This is a survey sheet, covering the principal types of sections used in industrial practice. Draw the required sections freehand, using an HB lead for visible lines and a *very sharp* F lead for section lines. Sketch over the half-toned lines to produce lines of the correct type and weight, converting hidden lines to visible lines where necessary, and adding any necessary lines to complete the sectional views. Space the section lines about 3 mm (.12") or more apart by eye, using the section lining for cast iron on all sections. Omit all hidden lines from the sectional views.

Sheet 42. Sectional Views. References: §§7.34, 9.1–9.7, 12.5, 13.17. Draw the indicated sectional views, using instruments. Use an F lead for all visible lines and a *very sharp* 2H lead for section lines. In all problems except Prob. 4, the location of the cutting plane is obvious, and it is unnecessary to show the cutting-plane line.

Sheet 43. Sectional Views. References: §§7.34, 9.1–9.8, 9.13. Draw the indicated sectional views, using instruments. Use an F lead for all visible lines and a *very sharp* 2H lead for section lines. In all problems except Prob. 4, the location of the cutting plane is obvious, and it is unnecessary to show the cutting-plane line.

Sheet 44. Sectional Views. References: §§7.34, 7.35, 9.1–9.4, 9.9, 9.12, 9.15, 12.5, 13.17. Draw the indicated sectional views, using instruments. Use an F lead for visible lines and a *very sharp* 2H lead for section lines. Add center lines. Omit finish marks unless assigned. In Space 1, include a revolved partial section in the front view to show the shape of the triangular rib. See Fig. 9.16 (m). In Space 2, fillets are 3 mm (.12") R.

Sheet 45. Sectional Views. References: §§5.22, 5.27, 9.9, 9.10, 9.16.

Space 1. Draw revolved sections as indicated, using break lines on each side of the octagonal section.

Space 2. Draw revolved sections, with "S" breaks on each side of sections. Inside diameter of the tube is 19 mm (.75").

Space 3. Draw removed sections, including all visible lines behind the cutting plane in each case, and all necessary center lines.

Sheet 46. Sectional Views. References: §§7.34, 9.1, 9.5, 12.5, 13.17. Draw sections as indicated, including visible lines beyond the sections. Add center lines. If assigned, add finish marks.

Sheet 47. Sectional Views. References: §§7.33, 7.34, 9.1, 9.5, 9.10, 9.12, 12.5, 13.17. Draw sections as indicated, including visible lines behind the cutting planes. Add center lines. If assigned, add finish marks.

Sheet 48. Sectional Views. References: §§7.30, 7.34, 9.1, 9.5, 9.11, 12.5, 13.17. Draw sections as indicated, including visible lines beyond the sections. Add center lines. If assigned, add finish marks.

Sheet 49. Sectional Views. References: §§7.29, 7.34, 9.11, 12.5, 13.17. Draw section as indicated. Add center lines. Show finish marks, if assigned, except for holes; these are understood to be finished.

Sheet 50. Sectional Views. References: §§7.33, 7.34, 9.1–9.7, 12.5, 13.17. Draw sections as indicated. Add center lines. If assigned, add finish marks.

Sheet 51. Sectional Views. References: §§7.33, 7.34, 9.1–9.5, 9.11, 12.5, 13.17. Draw sections as indicated, including all visible lines behind the cutting plane. Add center lines. If assigned, add finish marks.

Sheet 52. Auxiliary Views. References: §§6.2–6.8, 10.1–10.10, 10.16. Sketch the required auxiliary views as indicated. In Probs. 1 and 2, number the points in the auxiliary views to agree with the given views. Letter the numbers the size of the given numbers, without guide lines. Use numbers in Probs. 3 and 4 if needed or if assigned by instructor. Show reference planes or folding lines in all four problems. Include all hidden lines.

Sheet 53. Auxiliary Views. References: §§7.9, 7.33, 10.1, 10.12–10.14.

Space 1. First draw the auxiliary view of the entire object. This will show the true circular shape of the central portion. Then complete partial front view.

Space 2. Draw partial views only. The right end is semicircular. Draw break lines to indicate the limits of the partial views. Do not use a visible line of the view in place of a break line, and do not place a break line so as to coincide with a visible line.

Sheet 54. Auxiliary Views. References: §§7.33, 10.1–10.12, 13.14. Draw the required auxiliary views, using instruments. In Prob. 1 add numbers in the auxiliary view, making them the same height as those given. In the remaining problems use numbers if needed or if assigned. Show reference planes or folding lines in all three problems. In Prob. 4, surface A is inclined, §7.21. Include all hidden lines.

Sheet 55. Auxiliary Views. References: §§10.1–10.14, 13.17. Draw the required views as indicated, showing reference planes or folding lines in all three problems. Omit hidden lines in Prob. 1, but include them in Prob. 3. Add finish marks in Prob. 3.

Sheet 56. Auxiliary Views. References: §§10.1–10.10, 10.17. Draw the views and the section as indicated. Include all hidden lines in Prob. 1, and omit them in the auxiliary section of Prob. 2. In Prob. 1, number the corners of the oblique surface in all views, making the numbers about 1.5 mm (.06") high.

Sheet 57. Auxiliary Views. References: §§10.1–10.11, 13.14. Construct the views as indicated. In both problems measure the required angles with the protractor, and dimension the angles in degrees, Fig. 13.17 (e). Include all hidden lines.

Sheet 58. Auxiliary Views. References: §§10.1–10.11, 10.13, 13.14. Draw the required views, including all hidden lines. In Prob. 1 dimension the required angle in degrees, Fig. 13.17 (e).

Sheet 59. Auxiliary Views. References: §§10.11, 10.19–10.22. Draw the required views as indicated, including all hidden lines. In Prob. 2 measure the true angle between surfaces A and B with a protractor, and dimension the angle in degrees, Fig. 13.17 (e). Use reference-plane lines in Prob. 1 and folding lines in Prob. 2. In Prob. 2 space the primary auxiliary view 28 mm (1.12") from the given top view.

Sheet 60. Auxiliary Views. References: §§ 10.19–10.22, 10.24. Draw the required views as indicated. In Prob. 2 the upper end of the part is semicircular, as shown in the pictorial drawing.

Sheet 61. Auxiliary Views. References: §§10.19–10.24. Draw the required views as indicated, including all hidden lines. Draw primary auxiliary view 35 mm (1.40") from the front view. As seen in the secondary auxiliary view, all angles of the slot are 90°; that is, the slot has a rectangular right section. This is indicated in phantom section in the pictorial drawing.

Sheet 62. Revolutions. References: §§5.28, 5.29, 11.1–11.4. In the problems below, letter answers to the questions, using the guide lines provided. Show all hidden lines.

Space 1. At the left are given the front and top views of an object to be revolved. At the right the front view has been revolved counterclockwise. Draw the resulting top and right-side views.

Space 2. At the left are given the front and top views of an object to be revolved. At the right the top view has been revolved counterclockwise. Draw the resulting front and right-side views.

Space 3. At the left are given the front and top views of an object to be revolved. At the right draw the front, top, and right-side views as follows: Let the given front view be the right-side view in the new drawing, but revolve it so that surface A will appear true size in the new front view. Locate corner X at X_1. This is the point of view of the axis of revolution.

Space 4. At the left are given the front and top views of an object. At the right draw the front, top, and right-side views with the object revolved until the true angle between surfaces A and B appears in the right-side view. Dimension the angle in degrees, Fig. 13.17 (e).

Sheet 63. Revolutions. References: §§7.25, 11.1–11.10. In the problems below, show all hidden lines.

Space 1. At the left are given two views in which one revolution has been already accomplished. At the right the object is further revolved about a different axis. Draw the resulting front and right-side views. Remember the rule: *If lines are parallel on the object, they will be parallel in any view of the object.* In joining points, note which lines should be parallel and make sure they are so drawn by sliding a triangle, §2.21.

Space 2. At the right are shown the two views of the objects to reduced scale and in the original or unrevolved position. At the left the top view is shown full size and revolved clockwise. Draw the resulting front and right-side views, showing all construction. Number the corners if necessary (numbers 1.5 mm [.06"] high). Two simple blocks may be used as models to assist in visualization. For each block give the dimensions in order from the smallest to the largest, for example, 60 mm × 105 mm × 185 mm. If assigned, give decimal-inch dimensions.

Sheet 64. Revolutions. References: §§5.28, 5.29, 7.25, 11.5. In the problems below, show all hidden lines and number all corners.

Space 1. Draw the right-side view.

Space 2. As shown by the given front view, the object has been revolved clockwise about an axis perpendicular to the frontal plane of projection. Draw the resulting top and side views.

Space 3. As shown by the top view, the object has been further revolved, counterclockwise this time, about an axis perpendicular to the horizontal plane of projection. Draw the resulting front and side views.

Space 4. The object is to be further revolved, this time 15° clockwise about an axis perpendicular to the profile plane of projection. Draw the side view so that point **4** coincides with point **4** of the object. The side view must be an exact copy of the side view in Space 3, but revolved 15° clockwise. Draw the resulting top and front views.

Sheet 65. Revolutions. References: §§11.10, 11.13. The given figure **1–2–3–4–5–6** is the front face of a notched block 22 mm (.88") thick. Note that the face is bounded by two sets of parallel edges.

Space 1. Draw the right-side view of the surface.

Space 2. Revolve the face until either of the sets of parallel edges is true length in some view; that is, revolve either the top, front, or side view until one of the sets of parallel edges is vertical or horizontal. Then draw the other views in conformity.

Space 3. Continue the counterrevolution by revolving the surface until the other set of parallel edges is shown true length in some view, but keeping the first set of edges parallel to the plane to which it was revolved parallel in Space 2. Notice now that the two sets of edges will be parallel to two different planes and that the face does not show true size and shape in any view.

Space 4. Complete the counterrevolution by revolving the surface until both sets of edges are parallel to the *same* plane, in which position the face will be shown in its true size and shape and in an unrevolved position.

At any point in the counterrevolution where surface **1–2–3–4–5–6** shows as a line, the true depth of the object, 22 mm (.88"), may be set off perpendicular to it and the rest of the object drawn in all three views including all hidden lines. If this condition is not recognized in Space 3, you may easily complete the views in Space 4 and then complete the views in Spaces 3, 2, and 1 as required.

Sheet 66. Isometric Sketching. References: §§6.1–6.8, 6.12–6.14.

Problems 1 to 6. Sketch isometrics of given objects in spaces provided for each problem. Note the given starting point A.

Sheet 67. Isometric Drawing. References: §§18.5–18.18. Omit hidden lines in all problems. All "box construction" and other construction lines should be made lightly with a sharp 4H lead, and should not be erased. Darken all visible lines with a sharp F lead.

Spaces 1, 2, and 3. Draw isometric drawings of the objects shown, locating the corners at A and using the dividers to transfer distances from the views to the isometrics.

Space 4. Draw isometric drawing, locating the corner A at the point A. Use the scale to set off dimensions. Do not transfer distances with dividers, as the given drawing is not to scale. Show construction for the 30° angle.

Space 5. Complete the isometric drawing, using the information supplied in the reduced-scale drawing.

Space 6. Complete the isometric drawing, transferring measurements directly from the given views to the isometric with dividers. Draw the final curves with the aid of the irregular curve, §2.53.

Sheet 68. Isometric Drawing. References: §§18.5–18.18. Draw isometric drawings with instruments, locating them at points A as indicated. Omit hidden lines. Make all construction lines lightly with a 4H lead and do not erase. In particular, show complete construction for angles. Use a sharp F lead for the visible lines. In Prob. 2, part of the construction will fall outside the allotted space.

Sheet 69. Isometric Drawing. References and instructions same as for Sheet 68. In Prob. 2, part of the construction will fall outside the allotted space. Note that the isometric in Prob. 2 is to be drawn with reversed axes, Fig. 18.10, and is to be half-size.

Sheet 70. Isometric Drawing. References and instructions same as for Sheet 68.

Sheet 71. Isometric Drawing. References and instructions same as for Sheet 68. In addition, study §18.22. The problem in Space 1 is similar to Prob. 2 on Sheet 32 and Fig. 7.28 in the text. For Prob. 2 the approximate four-center ellipses of Fig. 18.24 may be used, but for the best results the ellipses should be drawn by the Orth method, Fig. 18.28, by a true-ellipse method, Fig. 18.22, or with the aid of an ellipse template, §18.21.

Sheet 72. Isometric Drawing. Reference: §18.24. Draw isometric sections, full size, of the objects shown, starting them at points A and B, as indicated. Draw the section lines about 3 mm (.12") or more apart, spaced by eye. Omit all hidden lines. Part of the construction may fall outside the allotted space.

Sheet 73. Oblique Sketching. Reference: §§6.3, 6.5, 6.15, 6.16. Using the method shown in Fig. 6.28, make oblique sketches of objects shown, using the starting corners indicated. Omit hidden lines and center lines. Make visible lines **black** so sketches will stand out clearly from the grids.

Sheet 74. Oblique Drawing. References: §§19.1–19.9. Draw oblique drawings full size with instruments, starting the problems at points A, as indicated. At points A also are shown the angles of the receding lines. In Prob. 1 transfer dimensions from the given views with dividers. On the remaining problems use the dimensions in the given views. Omit all hidden lines. Show all construction, especially for the points of tangency.

Sheet 75. Oblique Drawing. References: §§19.1–19.9. Draw oblique drawings full size with instruments, starting the problems at points A and B, as indicated. The angles of the receding lines are shown at these starting points. Omit all hidden lines. Show all construction, especially for the points of tangency. Study Fig. 19.13 carefully.

Sheet 76. Oblique Drawing. References and instructions same as for Sheet 75.

Sheet 77. Oblique Drawing. References: §§19.1–19.10. Draw oblique drawings full size with instruments, starting the problems at points A and B, as indicated. The angles of the receding lines are shown at these starting points.

Space 1. The upper right quarter is to be removed, and the cut surfaces are to be section-lined. Draw the section lines in the direction shown in Fig. 19.20, about 3 mm (.12") or more apart, spaced by eye. Omit all hidden lines. Show all construction, especially for the points of tangency.

Space 2. The right half of the object is to be removed, and the cut surfaces are to be section-lined. Omit all hidden lines and show all construction.

Sheet 78. Dimensioning. References: §§2.25, 2.27, 2.31, 4.16, 7.33, 7.34, 12.5, 13.1–13.11. Use the complete decimal dimensioning system with metric values. Add unidirectional dimensions freehand, Fig. 13.15 (a), spacing dimension lines approximately 10 mm (.40") from the views and 10 mm (.40") apart. If assigned, use decimal-inch dimensions.

Space 1. The small holes should be dimensioned similar to Fig. 13.36 (b). The large hole is bored. Give the diameter directly on the view, toleranced to two decimal places (three decimal places for decimal-inch dimensions), the minimum diameter being 19.05 mm (.750") and the tolerance .08 mm (.003"). Add finish marks, Fig. 13.20.

Space 2. The large hole is reamed, and should be dimensioned by note as shown in Fig. 7.40 (b). The minimum diameter is 44.44 mm (1.750"). Allow a tolerance of .13 mm (.005"). Give width of slot, not radii of ends. Add finish marks, Fig. 13.20.

Sheet 79. Dimensioning. References: §§2.25, 2.27, 2.31, 4.16, 7.33, 7.34, 12.5, 13.1–13.25, 13.29–13.31. Measure the views (from centers of lines), and dimension completely, using the complete decimal dimensioning system with metric values to the nearest 0.5 mm (.02"). If assigned, use decimal-inch dimensions. Notice the scales in each problem. Space dimensions uniformly 10 mm (.40") from the object and 10 mm (.40") apart. Use the unidirectional system, Fig. 13.15 (b). Draw light guide lines for all dimension figures and notes, §4.20. Use a sharp 2H lead for dimension lines, extension lines, and center lines; a sharp F lead for arrowheads and lettering. Give radii dimensions for the larger arcs, Fig. 13.19, and give notes covering the small fillets and rounds, §13.16. Use V-type finish marks in Spaces 1 and 2, Fig 13.20. In Space 2 give the angle in degrees, Fig. 13.17 (a). The large hole is reamed for an RC 6 fit, §14.8 and Appendix 5. Give the ream note in millimeters rounded off to two decimal places, Fig. 7.40 (b). *Caution:* Put dimensions in place before lettering notes; then choose the best open spots for the notes and letter them without crowding.

Sheet 80. Dimensioning. References: §§7.33–7.36, 12.3–12.5, 12.20, 13.1–13.31, 13.43, 14.1–14.7. Complete the detail drawing of the forged steel Coupler Block. First, complete the three views, then dimension the drawing and fill in the title strip, using the unidirectional complete decimal dimensioning system with metric values. If assigned, use decimal-inch dimensions. Two of these forgings are needed in each assembly of a Hydraulic Ram. Fillets and rounds are 3 mm (.12") R unless otherwise shown, §13.16. The drawing is half the size of the actual part, §2.31. Obtain dimensions by scaling the views. In addition to the surfaces shown finished (not added) in the pictorial drawing, the back and bottom surfaces are also finished. Give the angle of the inclined surface in degrees, Fig. 13.17 (a). The upper hole is reamed with .05 mm (.002") tolerance, the basic size being 19.05 mm (.750"). The lower hole is drilled, then counterbored on one end and spotfaced on the other. Use a single note covering all three operations, omitting the depth of the spotface. Space all dimensions 10 mm (.40") from the views and 10 mm (.40") apart.

Sheet 81. Dimensioning. References: §§7.33, 7.34, 12.3–12.5, 12.20, 13.1–13.31, 13.43, 14.1–14.7. Dimension fully, using the unidirectional combination dimensioning system. If assigned, use metric dimensions. Space dimensions .40" (10 mm) from the views and .40" (10 mm) apart. Include all finish marks. Give tolerances for reamed holes A and B to three decimal places (two places for metric dimensions). Hole A: Max. dia. = .375" (9.52 mm); tolerance = .005" (.13 mm). Hole B: Max. shaft dia., for hole = .625" (15.87 mm); allowance = .003" (.08 mm); tolerance = .002" (.05 mm). Tapped hole: No. 10 dia., Unified Coarse thread, Class 2B fit. See Fig. 15.19 (a). The small holes are drilled. For the chamfer, use note as in Fig. 13.46 (b).

Sheet 82. Dimensioning. References: §§13.1, 14.1–14.10. Give tolerance dimensions indicated, using the guide lines provided. If assigned, use metric values and round off to two decimal places in spaces 1–4 and three decimal places in spaces 5 and 6; see §14.1. Draw additional vertical or inclined guide lines, depending upon whether vertical or inclined lettering is required. The following letter-symbols are used: B = basic size, A = allowance, T = tolerance, TH = tolerance for hole, and TS = tolerance for shaft. Also Max. Int. = maximum interference, and Max. Clear. = maximum clearance. For Probs. 5 and 6, see § 13.8 and Appendixes 5–9.

Sheet 83. Dimensioning. References: §§7.33–7.36, 12.3–12.5, 12.20, 13.1–13.31, 13.43, 14.1–14.10. Dimension the Bracket fully. Include finish marks and notes. Use unidirectional combination system of dimensioning with metric values. If assigned, use decimal-inch dimensions. Space dimension lines 10 mm (.40") from the views and 10 mm (.40") apart. The two small holes require an allowance of 0.8 mm (.03") on the diameters for the M10 or 3/8" hexagon head cap screws. For the large hole, use a Class LC 8 fit, see §14.8 and Appendix 6. Note that the LC 8 fit also applies to the diameter of the Wheel Screw on Sheet 84. The keyway is for a No. 405 Woodruff key. Fig. 13.44 (p).

Sheet 84. Dimensioning. References: §§13.9, 13.20–13.27, 13.31, 14.1–14.10, 15.21. The Wheel Screw and Wheel fit with Bracket on Sheet 83 and are to be fully dimensioned in the manner used on Sheet 83. For the large hole in the Wheel, use a Class RC 7 fit; see Appendix 5. Note that the thread of the Wheel Screw also must pass through the large hole in the Bracket. Dimension the Wheel Screw mating dimensions accordingly. Provide an allowance of 0.08 mm (.003") and a tolerance of 0.05 mm (.002") for the width of the Wheel hub, calculated from the Wheel hub width as the basic size. The tolerances should permit the Wheel to turn freely on the Wheel Screw. The angular channel in the Wheel hub is a drilled hole for oil. The thread is 16 mm (or 5/8") diameter, metric coarse or Unified coarse, Fig. 15.19 (a), and is used with a lock washer and nut given in the standard parts list; see Fig. 15.31 (c). Complete the standard parts list with the size specifications for the hex nut and the lock washer in the spaces indicated. Locate the keyseat and by a note specify the key to be used, Fig. 13.48 (a).

If assigned, draw an assembly of the Truck Wheel with part identification numbers and a parts list, §§16.14, 16.20–16.22.

Sheet 85. Threads and Fasteners. References: §§ 15.1–15.21, 15.29–15.35. Draw light horizontal guide lines from the marks indicated and letter the answers in the spaces provided. Standard abbreviations may be used to avoid crowding; see Appendix 4.

Sheet 86. Threads. References: §§ 15.1–15.7, 15.11, 15.19–15.21. Draw detailed metric and Unified coarse threads, and letter thread notes as indicated. Chamfer the ends of the threads 45° to the depth of the thread. In Prob. 1 the threads are 42 mm dia. and should appear similar to those in Fig. 15.4 (a), but note that the thread will terminate on the left so that the last crest line will be just on or to the left of the limiting thread length dimension. In Prob. 2 the threads are 2.00" dia. Note that a neck is provided. The thread will simply run up to the side of the neck similar to that in Fig. 15.16. Refer to Appendix 15 for the number of threads per inch. Assume Classes 2A and 2B fits for the Unified threads. Complete the section lining after drawing the internal threads.

Sheet 87. Threads. References: §§15.3–15.7, 15.12, 15.21. Draw detailed square threads as indicated, and complete the section lining. For the number of threads per inch, see Appendix 22. A neck .20" (5 mm) wide is to be provided on the threaded shaft adjacent to the hub.

Sheet 88. Threads. References: §§15.3–15.7, 15.13, 15.21. Draw detailed Acme threads as indicated, and complete the section lining. For the number of threads per inch, see Appendix 22. A neck .20" (5 mm) wide is to be provided on the threaded shaft adjacent to the hub.

Sheet 89. Threads. References: §§15.3–15.7, 15.11, 15.19–15.21, Appendix 15. Using detailed representation, show the specified threads for the given sectioned assembly of the cylindrical Coupler. The internal thread of the Core is a through thread and the stud of the Piston Rod is engaged to the depth indicated. Complete the section lining and the thread-note leaders.

Sheet 90. Threads and Fasteners. References: §§15.7–15.10, 15.15–15.21, 15.23–15.28, Appendixes 18 and 27. Draw specified thread and fastener details, using schematic thread symbols in Probs. 1 and 3, and simplified thread symbols in Probs. 2 and 4, unless otherwise assigned. Complete the section lining and leaders where required. Chamfer both ends of threads 45° thread depth in Probs. 1 and 2, and the left ends of the external threads in Prob. 3.

Sheet 91. Fasteners. References: §§15.7–15.15, 15.22–15.24, 15.28–15.31, 15.34, Appendixes 18, 20, 23, and 27. Complete the sectioned assembly by drawing the specified fasteners, using simplified thread symbols unless otherwise assigned. Complete the required section lining of the sectioned assembly of a Hydraulic Pump. The slotted Rotor is keyed to the shaft and when it is rotated, the centrifugal force actuates six nylon or neoprene rollers to force fluid through an exhaust port that is not shown.

Sheet 92. Springs. References §§15.37, 15.38.

Prob. 1. A pictorial of a round wire compression spring having 5 total coils is shown in Fig. 15.50 (a). The steps in drawing a detailed elevation of this spring are shown at (d), (e), and (g). In Prob. 1, draw a similar compression spring in elevation. OD = 1.75" (44.45 mm); No. 3 (.2437") W&M gage round steel wire; free length = 2.25" (57.15 mm); 6 total coils; LH wound; and ends closed ground.

Prob. 2. The construction for a schematic elevation view of a compression spring having 6 total coils in shown in Fig. 15.49 (a). In Prob.2, draw a similar spring but with 8 total coils. ID = 1.62" (41.15 mm), free length = 1.75" (44.45 mm); LH wound; and ends closed and ground. Make lines same weight as visible lines.

Prob. 3. Draw a schematic elevation view of an extension spring with 7 active coils, similar to Fig. 15.49 (c). OD = .62" (15.75 mm); pitch = .25" (6.35 mm); LH wound; and loop ends. Center left loop at point A. Make lines same weight as visible lines.

Prob. 4. Draw a detailed compression spring in section with 5 total coils, Fig. 15.50 (f). ID = 1.25" (31.75 mm); .25" (6.35 mm) square wire; free length = 2.25" (57.15 mm); LH wound; and ends closed and ground. Do not erase construction. See Fig. 15.44 (b).

Prob. 5. Draw a detailed spring in section, compressed between A and B, with 7½ total coils, similar to Fig. 15.50 (b). Pitch diameter = 1.25" (31.75 mm); No. 5 (.2070") W&M gage round steel wire; free length = 2.69" (68.33 mm); LH wound; and ends closed and ground. Do not show hidden lines for portion of the spring that falls behind the hub above point B, but show all visible lines in the sectional view. Add missing visible lines in the assembly adjacent to the spring. Use steel section lining.

Sheet 93. Graphs. References: §§28.1, 28.15, 28.16, 28.19, 28.20.

Space 1. Construct a pie chart in the given circular area showing passenger car production in the United States during 1988 from the table on the next page. Letter the chart appropriately, including the title PASSENGER CAR PRODUCTION and subtitle USA—1988.

Manufacturer	Units	Percent
Chrysler Corp.	1,072,845	15.1
Ford Motor Co.	1,805,741	25.4
General Motors	3,501,124	49.2
Honda	366,354	5.2
Others	364,654	5.1

Space 2. Construct a horizontal bar chart comparing the average annual increase in output per man-hour in manufacturing for several countries for 1960–1973. Arrange the bars in order with the longest at the top. Make the bars .20" (5.08 mm) wide and space them .10" (2.54 mm) apart. Shade them with thin dark parallel lines as in Fig. 28.31 (a). Complete the fine dark grid lines but do not draw across the bars. Letter the names of the countries in a neat column at the left of the vertical axis and opposite the corresponding bar.

Country	Percent Gain
Japan	11.5
Netherlands	8.5
Sweden	8.1
Belgium	7.45
Italy	7.3
France	7.0
West Germany	6.8
Switzerland	6.25
Canada	5.3
United Kingdom	5.0
United States	4.35

Sheet 94. Graphs. References: §§28.1–28.7. Using the data given in the table below, complete the graph showing the compression curves for individual pieces of timber and concrete. The loads, or stresses, are given in terms of pounds per square inch, and the unit strains for each piece are given as fractional parts of the original lengths, e.g., .0003" per inch or .0003' per foot. Plot the points for the values listed and draw the curves. Label the curves in an appropriate manner.

Stress	Strain	
lb/in.^{2} (psi)	Timber	Concrete
500	.00033	.0002
750	—	.0003
1000	.00067	.00045
1250	—	.0006
1500	.0010	.00082
1750	—	.0011
2000	.0013	.0018
2500	.0017	—
3000	.0020	—

Sheet 95. Graphical Algebra. References: §§2.54, 4.21–4.23, 31.1.

Probs. 1 and 2. Plot the two curves or simultaneous equations as indicated on the given axes. Determine graphically the solutions for the equations. Show on your drawing a check of the solution values in the given equations. Label the solutions and the curves.

Sheet 96. Graphical Algebra. References: §§2.54, 4.21–4.23, 31.1.

Prob. 1. Determine the solutions graphically by plotting on the given axes. Show on your drawing a check of the solution values in the given equations. Label the solutions and the curve(s).

Prob. 2. Plot the two curves or simultaneous equations as indicated on the given axes. Determine graphically the solutions and roots for the equations. Show on your drawing a check of the found values in the equations. Label the solutions, roots, and curves.

Sheet 97. Empirical Equations. References: §§28.1–28.10, 30.1–30.7. Plot the following data four times on the layout given, using (1) linear scales for both variables, (2) a linear scale for x and a logarithmic scale for y, (3) logarithmic scale for x and linear scale for y, and (4) logarithmic scales for both. One of these plots should be a straight line. From it determine the equation that relates the variables.

x	y
.8	4.2
1.4	5.4
2.1	7.2
2.8	9.7
3.7	14.1
4.5	19.6
5.5	29.8

Sheet 98. Nomographs. References: §§29.1, 29.8, 29.14, 29.15.

Prob. 1. Construct a natural parallel-scale nomograph or alignment chart for the equation $x + y - 2 = z$. Space the scales 3.0" (76.2 mm) apart. Make all scales the same length as the given x-scale. Match height of scale numbers to given numbers and identify each scale. Check the diagram by substituting the following values: $x = 0, y = 8; x = 12, y = 12$ and the appropriate z-scale value in the equation. Show calculations on the drawing.

Prob. 2. Construct a natural scale N-chart for the Ohm's law equation, $V = IR$, where V = voltage, I = current (scale range 0–10 amperes), and R = resistance (scale range 0–100 ohms). Match scale numerals to given numerals and label all scales. Check the chart by substituting these values: 10 amperes, 30 ohms; 2.5 amperes, 80 ohms, and the appropriate voltage values in the equation. Show the calculations on the drawing.

Sheet 99. Nomograph. References: §§ 29.1-29.5, 29.10, 29.11. Using the given layout, design a nomograph for solving the values of P in the equation $P = IR$. The range of I is to be .03–.8 and the range of R is to be 250–1250.

Give the scale modulus for each scale. Find the location and scale modulus of the center scale graphically unless otherwise assigned. Utilize the entire lengths of the given solid lines for the I and R scales.

Sheet 100. Different Calculus. References: §§31.2–31.11. The curve on the given x- and y-axes represents the distance from the starting point a body has moved, plotted against time. Differentiate this distance curve on the given x'- and y'-axes to produce a first derivative curve that will give the velocity of the object at any instant. Use the slope law method of §31.5 unless otherwise assigned.

Establish the base of each slope triangle as a value of 10 on the x-axis. The tangent of the slope angle is then one-tenth the value of the ordinate on the y-axis. Therefore, if the scale on the y'-axis is expanded in a 10:1 ratio to the y-axis, the ordinate of the slope triangle may be transferred directly to the construction for the first derivative.

Sheet 101. Integral Calculus. References: §§31.2, 31.12–31.19. The given curve of volume versus pressure is characteristic of data often encountered in science and technology. The integral in this case is the work curve (ft-lb vs. ft) and is equivalent to the area under the given curve. The work units are foot-pounds.

Determine the integral curve, using the area law method of §31.13 unless otherwise assigned. Add the appropriate labeling to the given x- and y-axes. Crosshatch the areas used to establish the mean ordinates. Show lengths of mean ordinates and summations of the ordinates used to establish the integral curve. Also show how one point on the integral curve was established.

Sheet 102. Engineering Design. References: §§31.12–31.19. In the upper half of the sheet is shown a plat of a 6.8-acre tract of land. It is bounded on the east, south, and west by roads and on the north by a stream. By graphical integration, find a line that represents its area. From this determine and record the scale of the map. The developer wishes to widen the street on the south by donating .3 acre to the municipality. He wishes to subdivide the remaining property into five lots of one acre each and one lot, the westernmost one, of 1½ acres, the dividing lines to be due north and south. Lay out all of the property lines (640 acres = 1 square mile).

In the lower half of the sheet is shown the cross section of a specially designed cylindrical 87-gallon gasoline tank drawn to the scale 1" = 1'–0. It was designed to fit the void in a fuselage. When in place, A will be the top and B the bottom of the tank. Using B as the origin and BA as the x-axis, integrate to find the area of the cross section. From this determine the length of the tank. Lay off along MN a measuring stick, graduated in increments of 10 gallons each, which may be inserted at A to determine the amount of gasoline in the tank. (*Hint:* First subtract graphically curve BDA from curve BCA to get a curve representing the tank's breadth at any level. Integrate this from B for the area curve.)

If assigned, metric units may be used. See Appendix 31.

Sheet 103. Computer-Aided Drafting: Terms and Descriptions. References: Chapters 3 and 8, Appendix 3. Some terms related to computer graphics are given in the table. A list of descriptions for these terms is given on the right. Find the matching description for each term and enter its letter identifier in the table.

Sheet 104. Computer-Aided Drafting: Two-Dimensional Coordinate Plot. Reference: Chapter 8.

Space 1. Digitize the single-view drawing by defining the X and Y coordinates of the indicated points and fill in the given table. Point A is the origin with values X and Y equal to zero. Consider each division of the grid as 1 unit. Keep in mind that any X values to the left of the origin and any Y values below the origin are negative.

Space 2. From the X and Y coordinate data given in the table, plot all points on the grid and complete the drawing. Point A is the origin. Consider each division of the grid as 1 unit.

Sheet 105. Computer-Aided Drafting: Three-Dimensional Coordinate Plot. Reference: Chapter 8. In drawing an image, the actions of the pen are Move and Draw.

Move: The pen moves from its present position to new X, Y, and Z coordinates specified. A line is not drawn. Numeral 0 is used to indicate Move action.

Draw: A line is drawn from the present pen position to new X, Y, and Z coordinates specified. Numeral 1 is used to indicate Draw action.

Space 1. Determine X, Y, and Z coordinates for all the points of the object. Complete the table for drawing the object, starting with point A. Coordinates X, Y, and Z are positioned as indicated by the arrows, with point A as origin. Try to use a minimum number of Move actions.

Space 2. According to the data shown in the table, draw the object on the grids provided. Coordinates X, Y, and Z are positioned as indicated by the arrows, with point A as origin.

Sheet 106. Computer-Aided Drafting: Menu Usage. Reference: Chapter 8. The drawing shows the front view of a Bracket that is to be generated on a graphics terminal. The numbers 1 through 21 refer to graphic entities that make up the drawing. Available menu commands for generating entities are given on the right. Complete the table by determining the menu commands to generate the entities. Enter the letter identifiers (A, B, etc.) of menu selections in the table.

Sheet 107. Computer-Aided Drafting: Coordinate Systems. Reference: Chapter 8. Using the given descriptions for VIEW COORDINATES and WORLD COORDINATES, complete the tables for the front and right-side views of the object. Point number 1 is considered as the origin. Each grid division is equal to 1 unit.

Sheet 108. Detail Drawings. References: Chapters 6–10, 13–15, §16.8. Draw or sketch the necessary views of the object assigned. Select appropriate scale and sheet size. Dimension completely using metric or decimal-inch dimensions as assigned.

Alternate Assignment: Using a CAD system, produce a hard-copy multiview drawing of the problem assigned. Dimension completely.

Sheet 109. Detail Drawings. References: Chapters 6–10, 13–15, §16.8. Draw or sketch the necessary views of the object assigned. Select appropriate scale and sheet size. Dimension completely using metric or decimal-inch dimensions as assigned.

Alternate Assignment: Using a CAD system, produce a hard-copy multiview drawing of the problem assigned. Dimension completely.

Before starting to letter, draw vertical guide lines entire height of sheet.

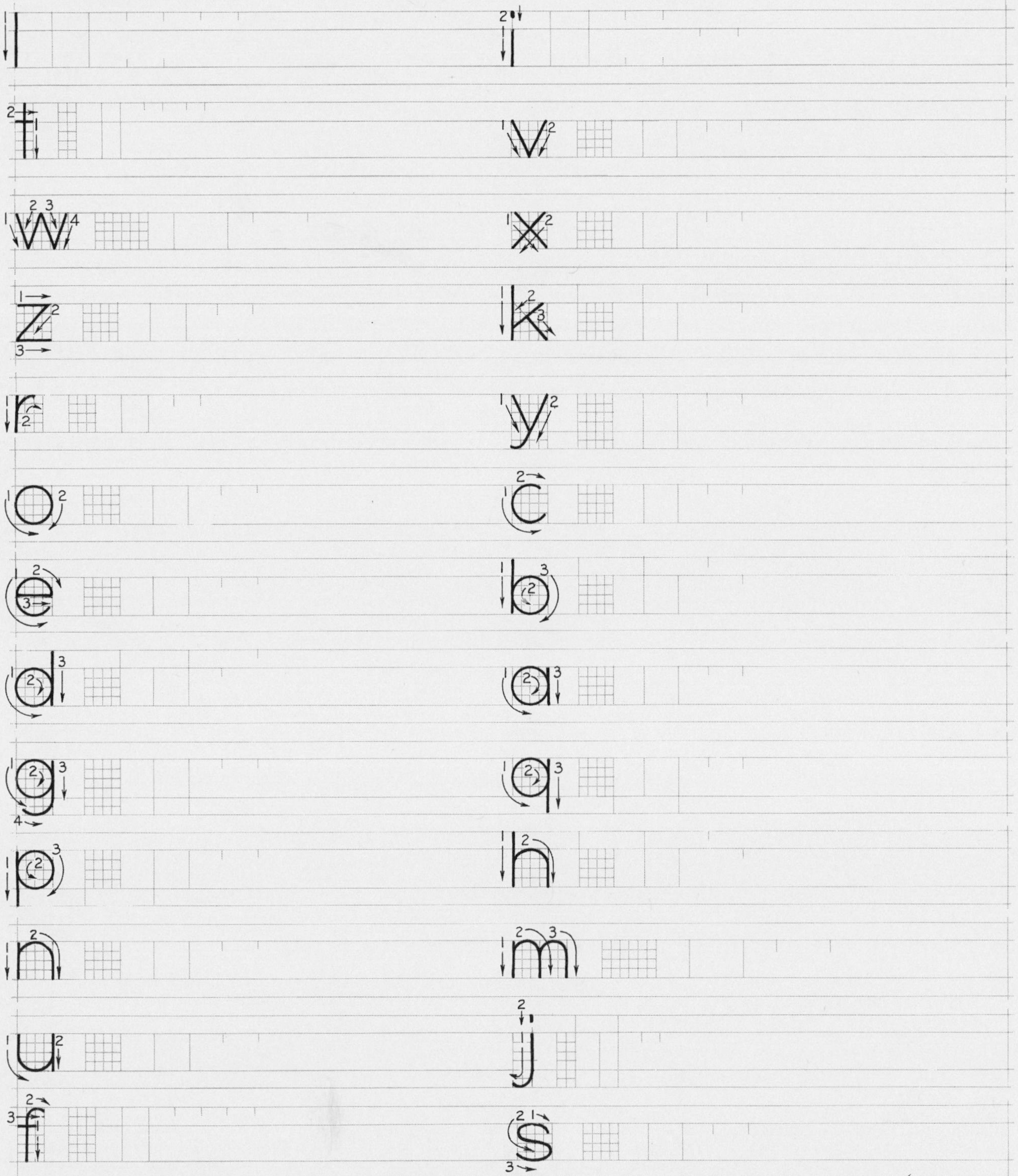

zfrskjqxawugvenptohcybmidlg

This method of executing letters by a systematic series of strokes

SECTION	DRAWN BY:		SHEET 5

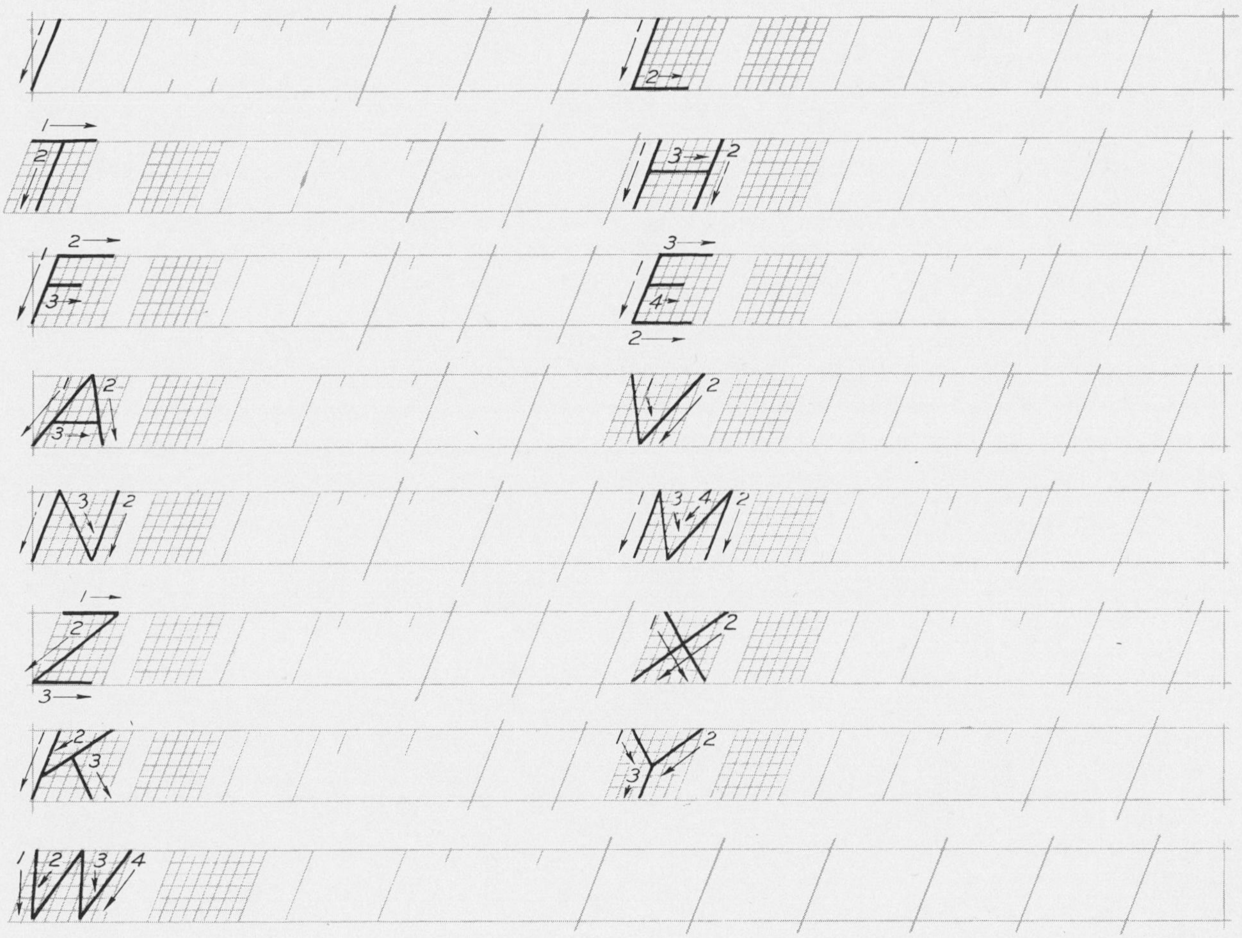

VENTILATE INITIAL ETHYLENE

ANALYZE ALL-METAL MAXIM

FEMININE VANITY HALF FLIT

KNIFE ELIMINATE FELT WHITE

INANIMATE ZAYIN NINE TENT

SECTION	DRAWN BY:		SHEET 6

File number

O
C
Q
G
J
U
D
P
R
B
S
&

CURRICULUM JUSTICE BRIDE

GOUGE ORBIT FORECASTER

GYROCOMPASS PROCEDURE

QUADRATIC GROINS BRIDGE

BOISE & SONS PACKING CO.

COMPACT AND E X P A N D E D

SECTION	DRAWN BY:		SHEET 7

1 2

3 4

5 6

7 8

9 0

$\frac{1}{2}$ $\frac{3}{4}$ .50 .75

$\frac{5}{8}$ $\frac{9}{16}$ .62 .56

89 3.75 8926 43.59 2067

$\frac{3}{16}$ $\frac{19}{32}$ $\frac{27}{64}$ $\frac{5}{8}$ $\frac{15}{16}$ $\frac{1}{4}$ $\frac{5}{64}$ $\frac{1}{2}$ $\frac{31}{32}$ $\frac{5}{8}$ $\frac{9}{16}$ $\frac{3}{4}$ $\frac{13}{16}$ $\frac{29}{32}$

1937 8172 3206 9.514 7.231 3958 2.605 4793 2561

$5\frac{3}{8}$ $9\frac{9}{16}$ $8\frac{3}{4}$ $2\frac{5}{32}$ $20\frac{3}{8}$ $1\frac{7}{8}$ $39\frac{5}{64}$ $\frac{2.500}{2.498}$ $\frac{1.724}{1.720}$ $\frac{5.5898}{5.5898}$ $\frac{7.50}{7.50}$ $\frac{25.40}{25.20}$

SECTION	DRAWN BY:		SHEET 8

Before starting to letter, draw inclined guide lines entire height of sheet.

INCLINED GUIDE LINES, MADE INDE-

PENDENTLY OF THE WIDTHS OF THE

LETTERS AND SPACES, WILL AID IN

GETTING A CORRECT AND UNIFORM

SLANT FOR INCLINED LETTERS. THE

INCLINATION VARIES FROM 60° TO 75°. THE SPECIAL

SLOT IN THE BRADDOCK TRIANGLE IS CONSTRUCTED

AT A $67\frac{1}{2}°$ INCLINATION, AND GUIDE LINES SHOULD BE

DRAWN ALONG THE SIDE OF THE SLOT TO DETERMINE

THE INCLINATION OF THE LETTERS. THE CLEAR SPACE

BETWEEN SUCCESSIVE LINES OF LETTERS IS USUALLY

EQUAL TO TWO-THIRDS THE HEIGHT OF THE LETTERS.

SECTION	DRAWN BY:		SHEET 9

Before starting to letter, draw inclined guide lines entire height of sheet.

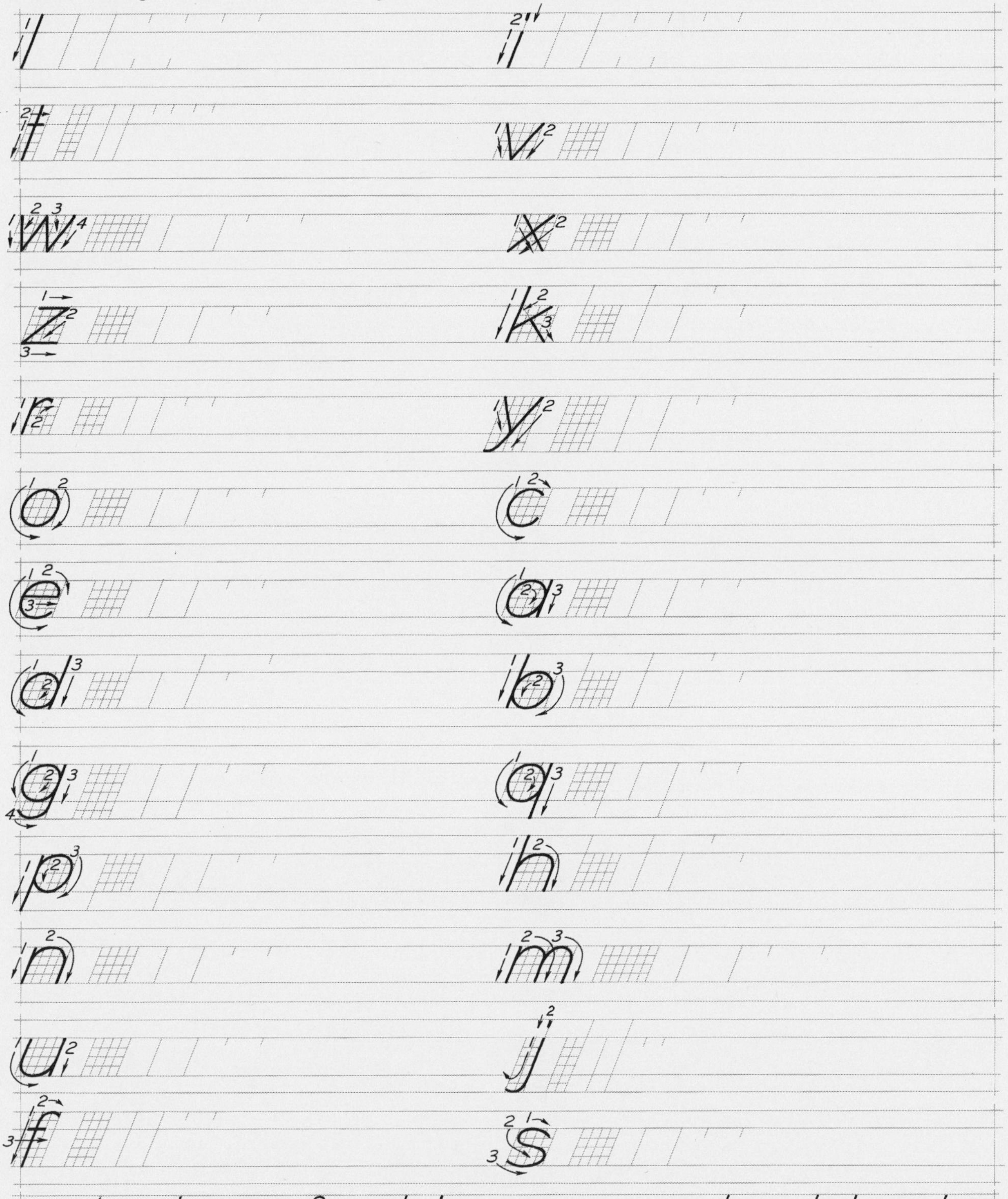

vehptugzfrskjqxawncybmidoplw

their major axes sloping at an angle of forty-five degrees with

SECTION	DRAWN BY:		SHEET 10

USE OF T-SQUARE AND TRIANGLES

1

2

3

4

5

6

SECTION	DRAWN BY:		SHEET 12

GEOMETRY OF TECHNICAL DRAWING – PARABOLA AND HYPERBOLA

(1) DRAW PARABOLA WITH F AS FOCUS AND AB AS DIRECTRIX.

A
F
B

(2) DRAW PARABOLA THROUGH C, D AND E, WITH E AS THE VERTEX.

C
E
D

(3) DRAW HYPERBOLA WITH F AND F' AS FOCI AND AB AS TRANSVERSE AXIS.

F
A
B
F'

(4) DRAW EQUILATERAL HYPERBOLA WITH GIVEN LINES AS ASYMPTOTES, THE CURVE PASSING THROUGH POINT P.

P

SECTION	DRAWN BY:		SHEET 22

MULTIVIEW PROJECTION – OPTIONAL VIEWS

(1)

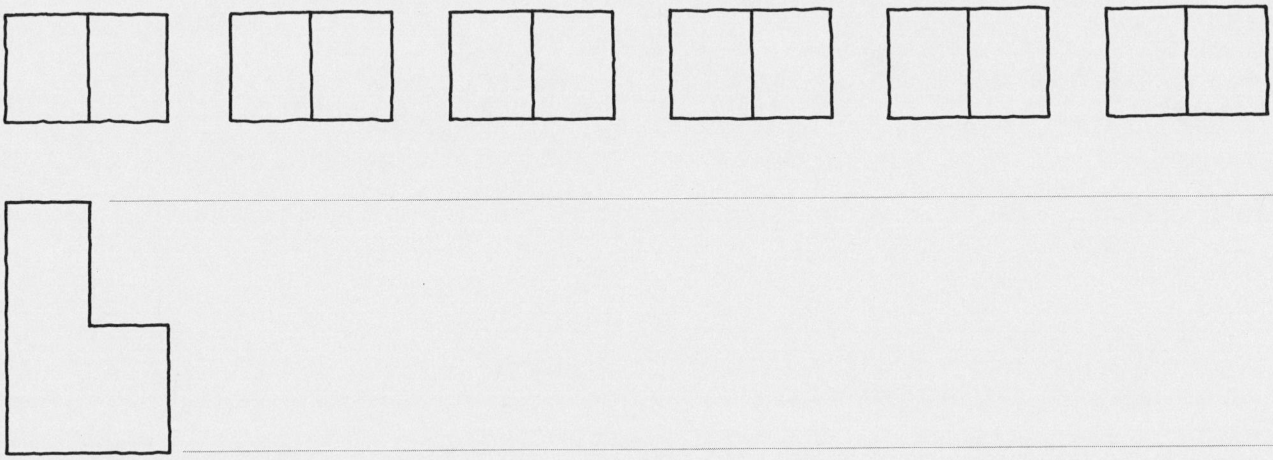

EXAMPLE

DESIGN DIFFERENT FRONT VIEWS WHICH WILL AGREE WITH GIVEN TOP VIEWS.
MAKE ALL FRONT VIEWS SAME HEIGHT.

(2)

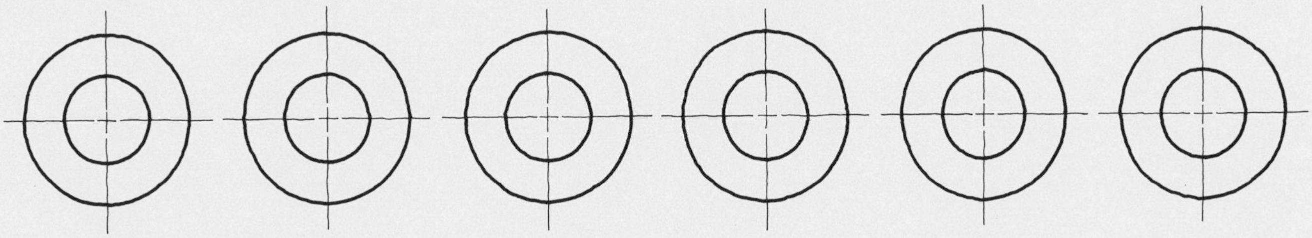

DESIGN DIFFERENT FRONT VIEWS WHICH WILL AGREE WITH GIVEN TOP VIEWS.
MAKE ALL FRONT VIEWS SAME HEIGHT. SHOW ALL CENTER LINES.

(3)

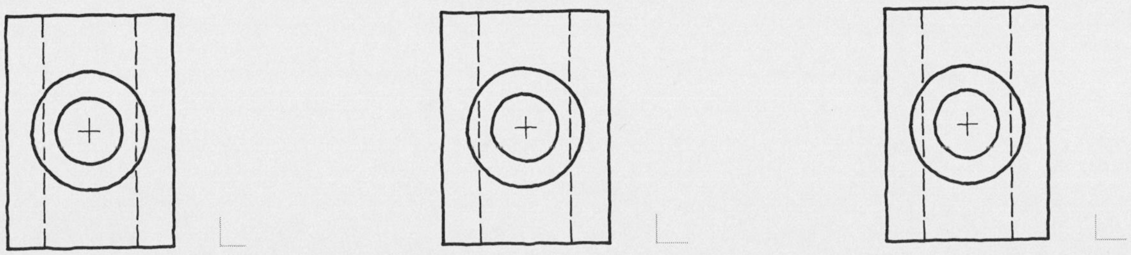

DESIGN DIFFERENT TOP AND RIGHT SIDE VIEWS WHICH WILL AGREE WITH GIVEN FRONT VIEWS.
SHOW ALL CENTER LINES.

SECTION	DRAWN BY:		SHEET 27

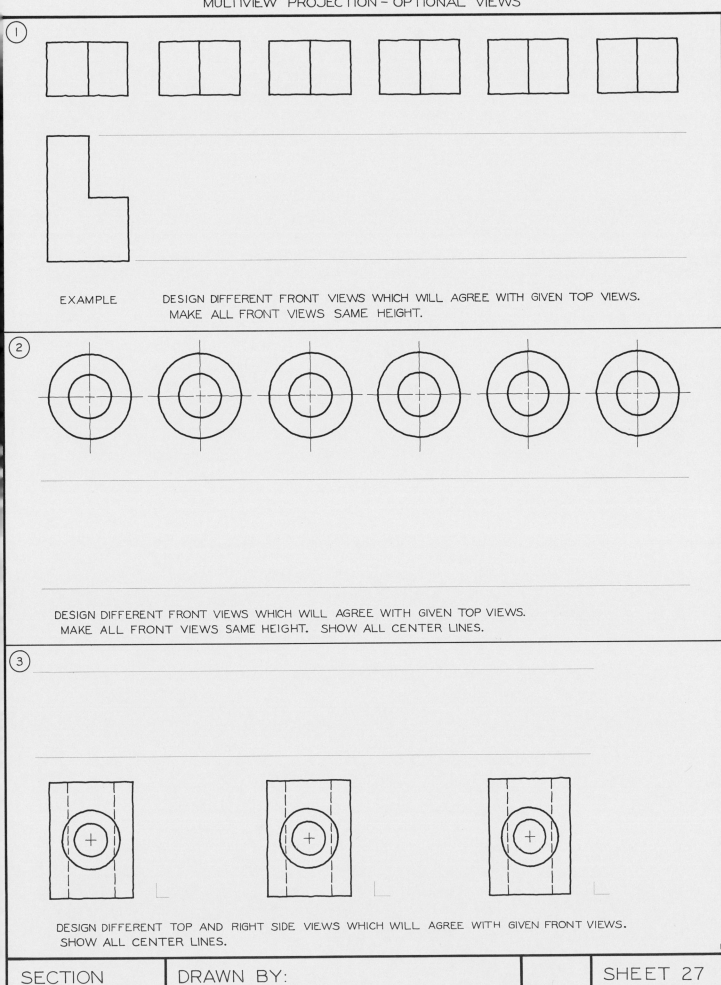

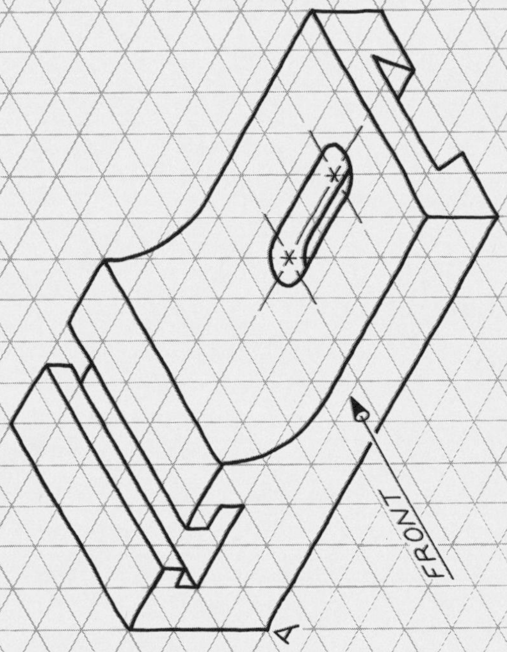
FRONT
A

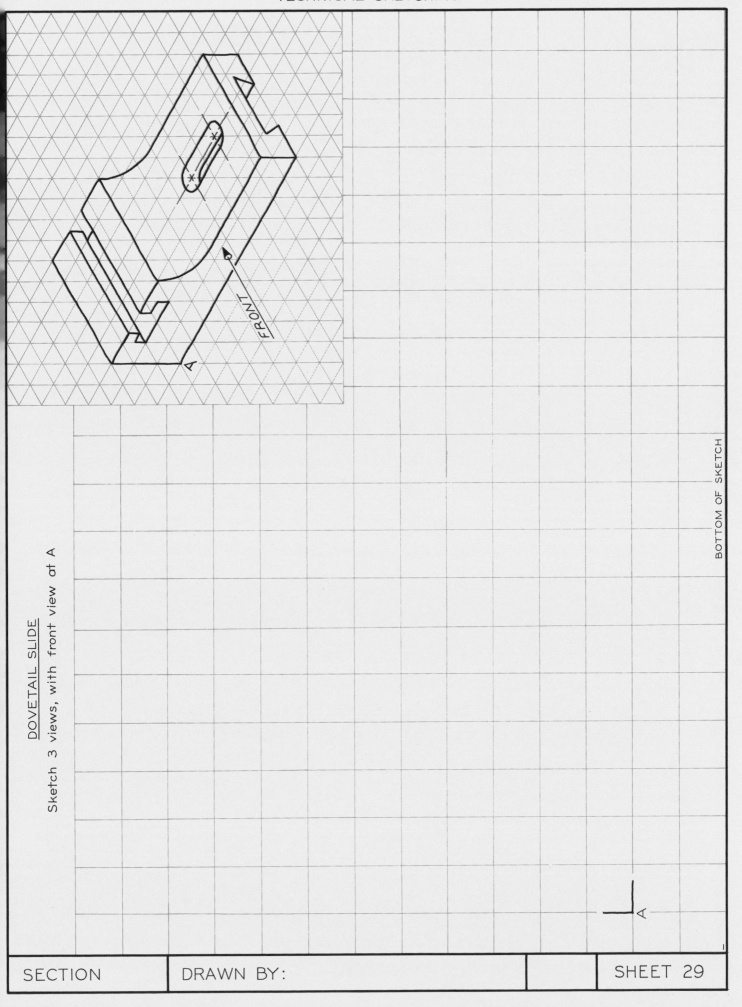

DOVETAIL SLIDE

Sketch 3 views, with front view at A

BOTTOM OF SKETCH

A

SECTION	DRAWN BY:		SHEET 29

MISSING VIEWS AND LINES

(1) LOCATOR

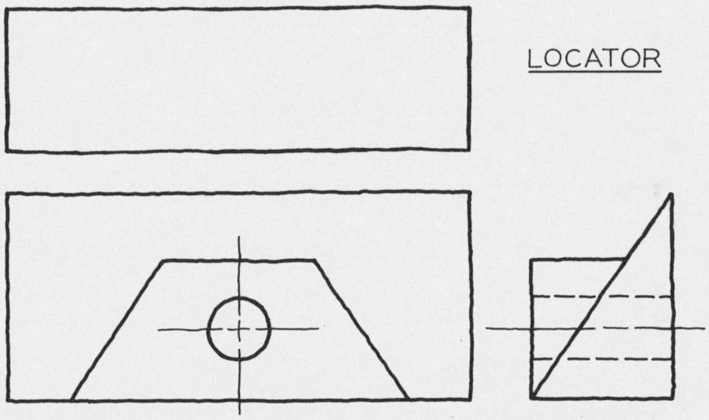

How many normal surfaces? ___ Inclined surfaces? ___

(2) WEDGE

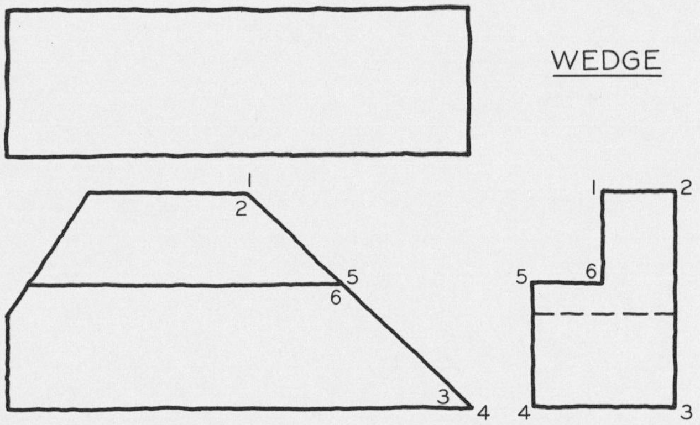
1
2
5
6
3
4
1
2
6
5
4
3

(3) SLIDE

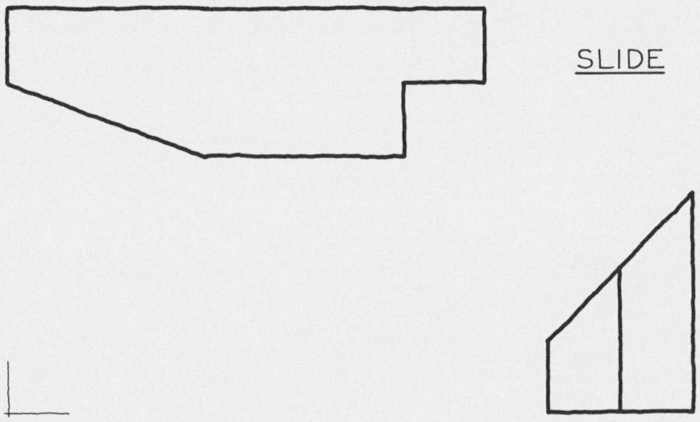

(4) FINGER

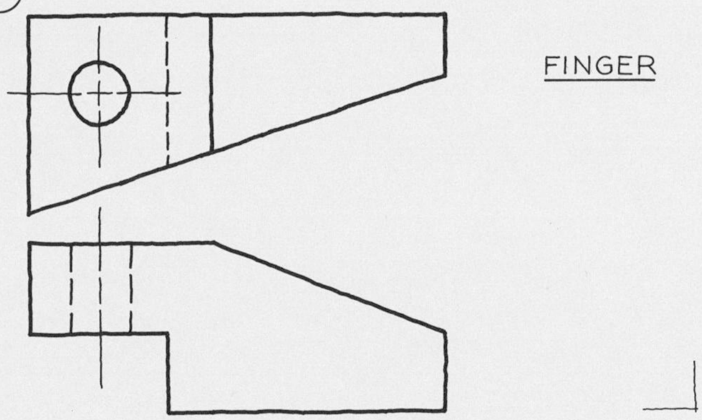

(5) ANGLE GUIDE

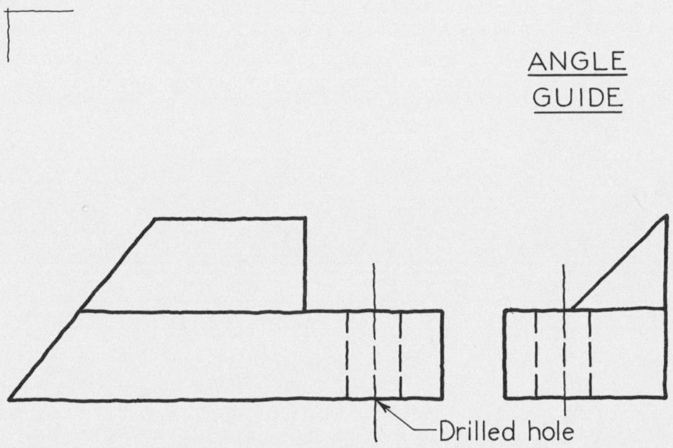
Drilled hole

(6) BEVEL BLOCK

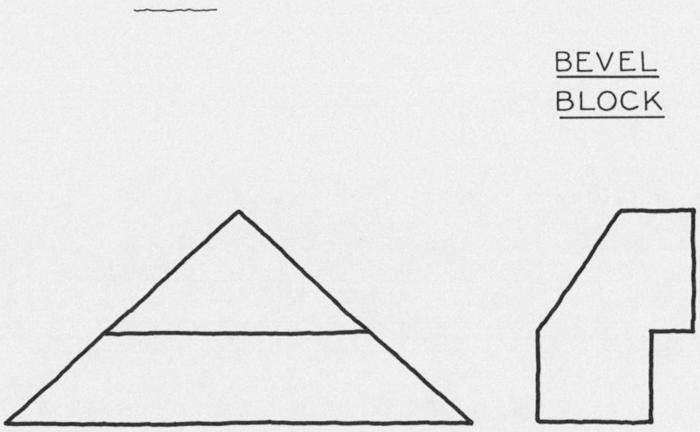

(7) V-BLOCK

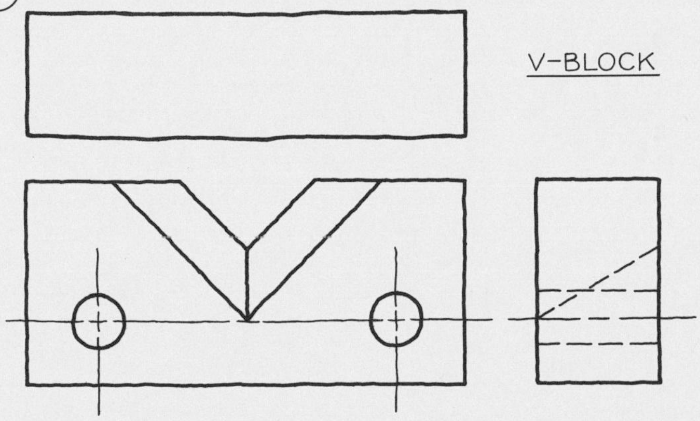

(8) TERMINAL

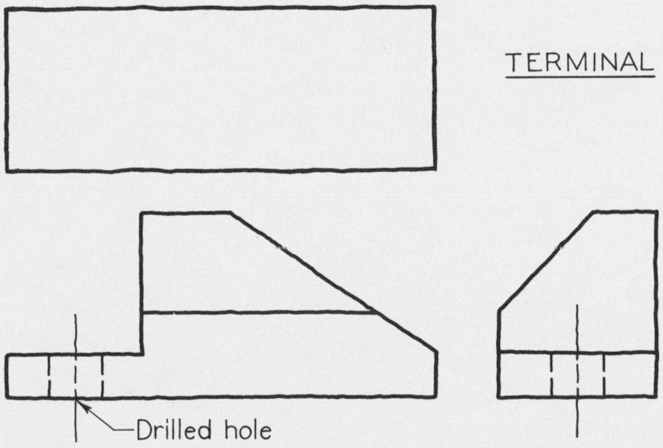
Drilled hole

SECTION	DRAWN BY:		SHEET 30

(1)

OFFSET
GUIDE

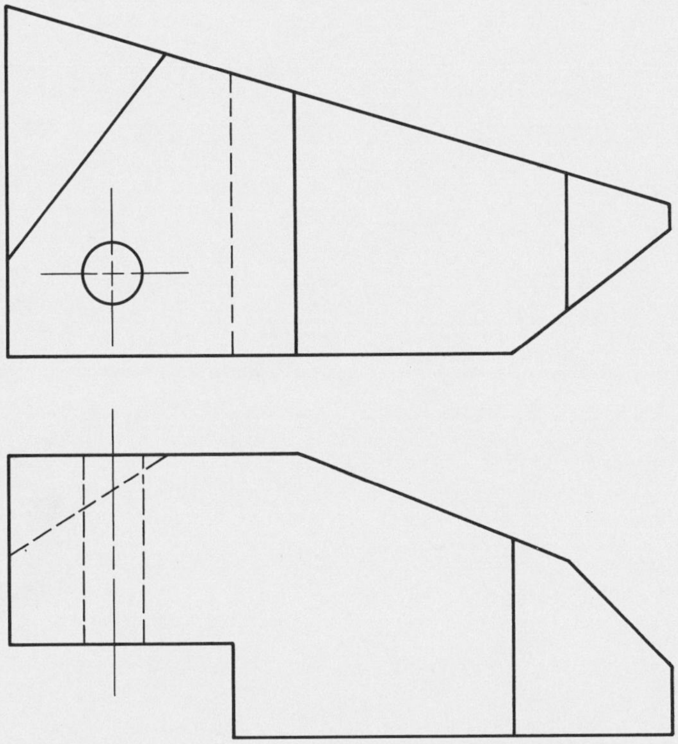

VIEW HERE

(2)

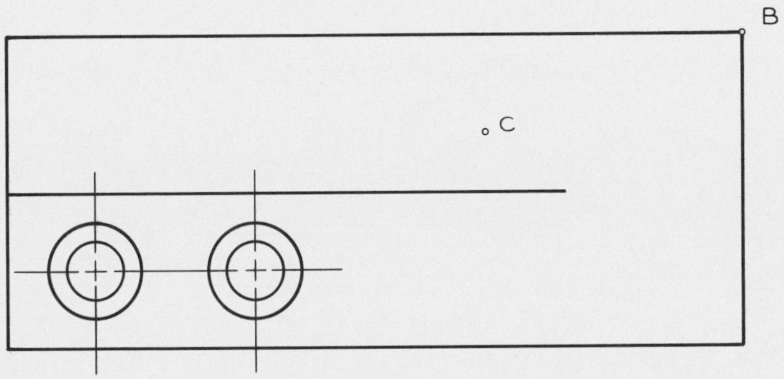
B
C

FIXTURE
BLOCK

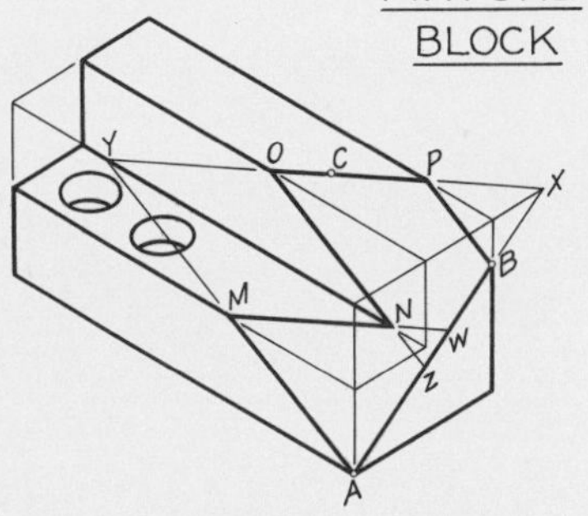
Y
O
C
P
X
B
M
N
W
Z
A

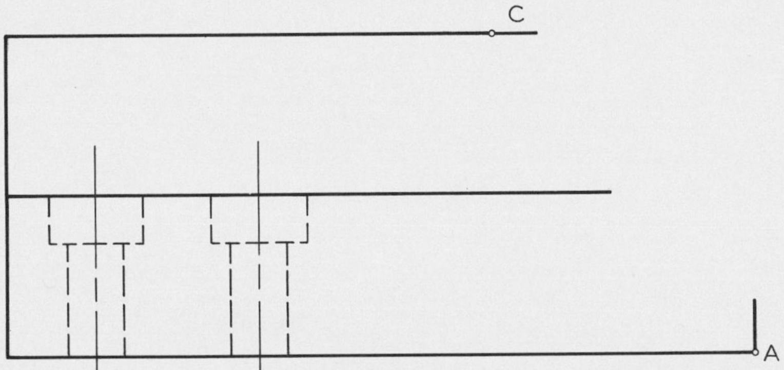
C
A

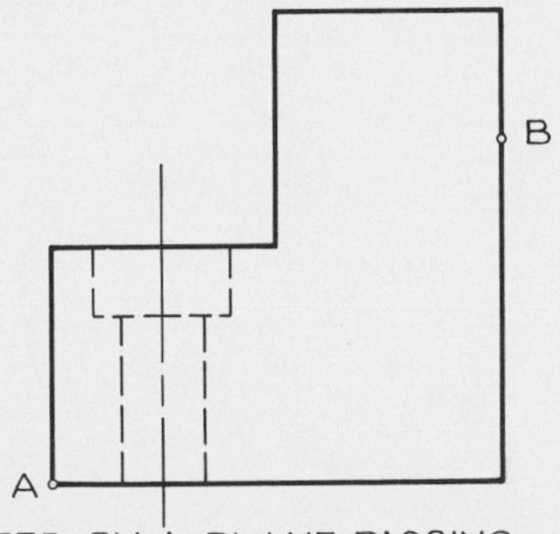
B
A

COMPLETE THE THREE VIEWS, THE OBJECT BEING TERMINATED BY A PLANE PASSING THROUGH POINTS A, B, AND C.

SECTION	DRAWN BY:		SHEET 32

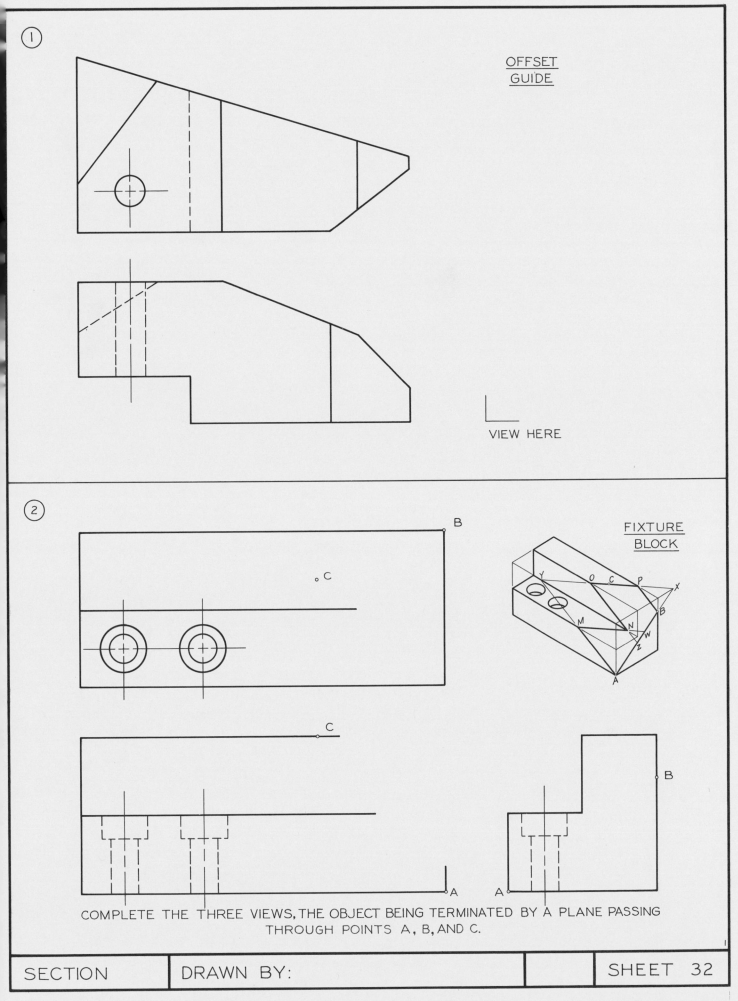

(1)

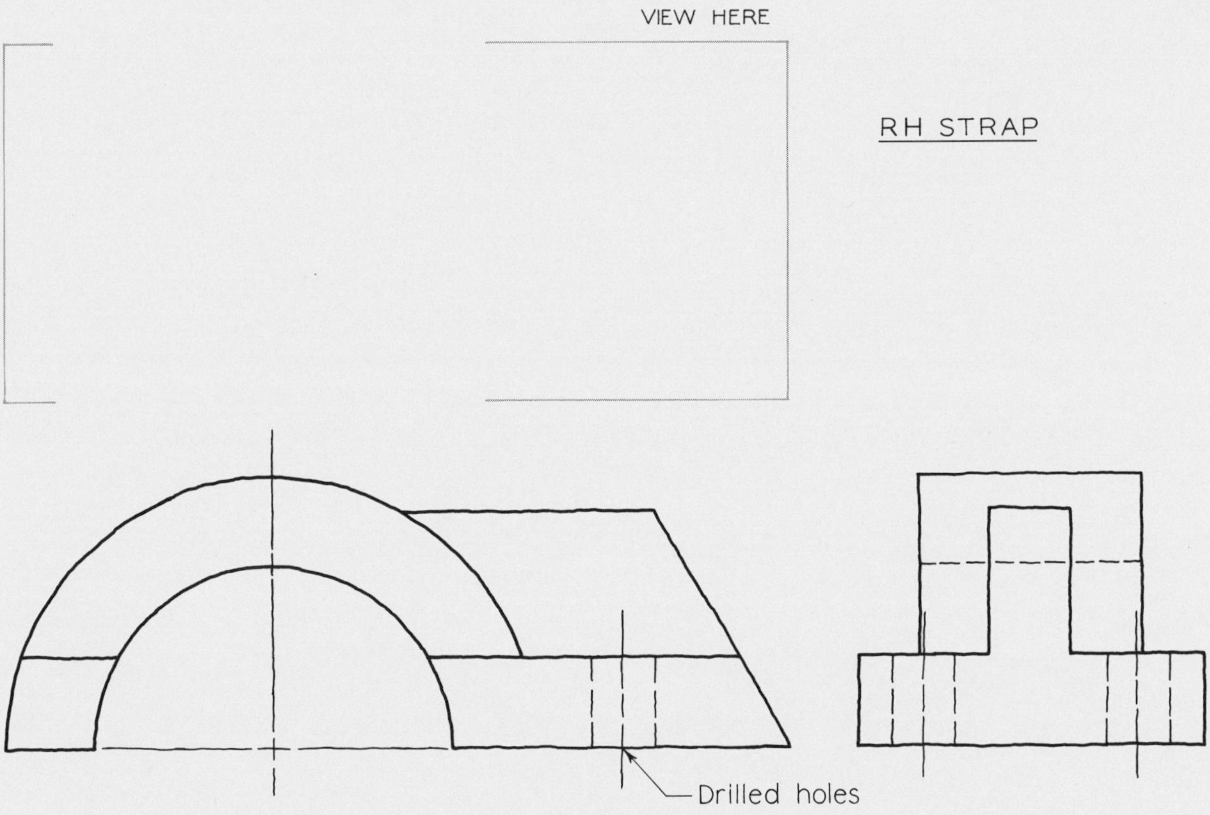
VIEW HERE
Drilled holes

RH STRAP

(2)

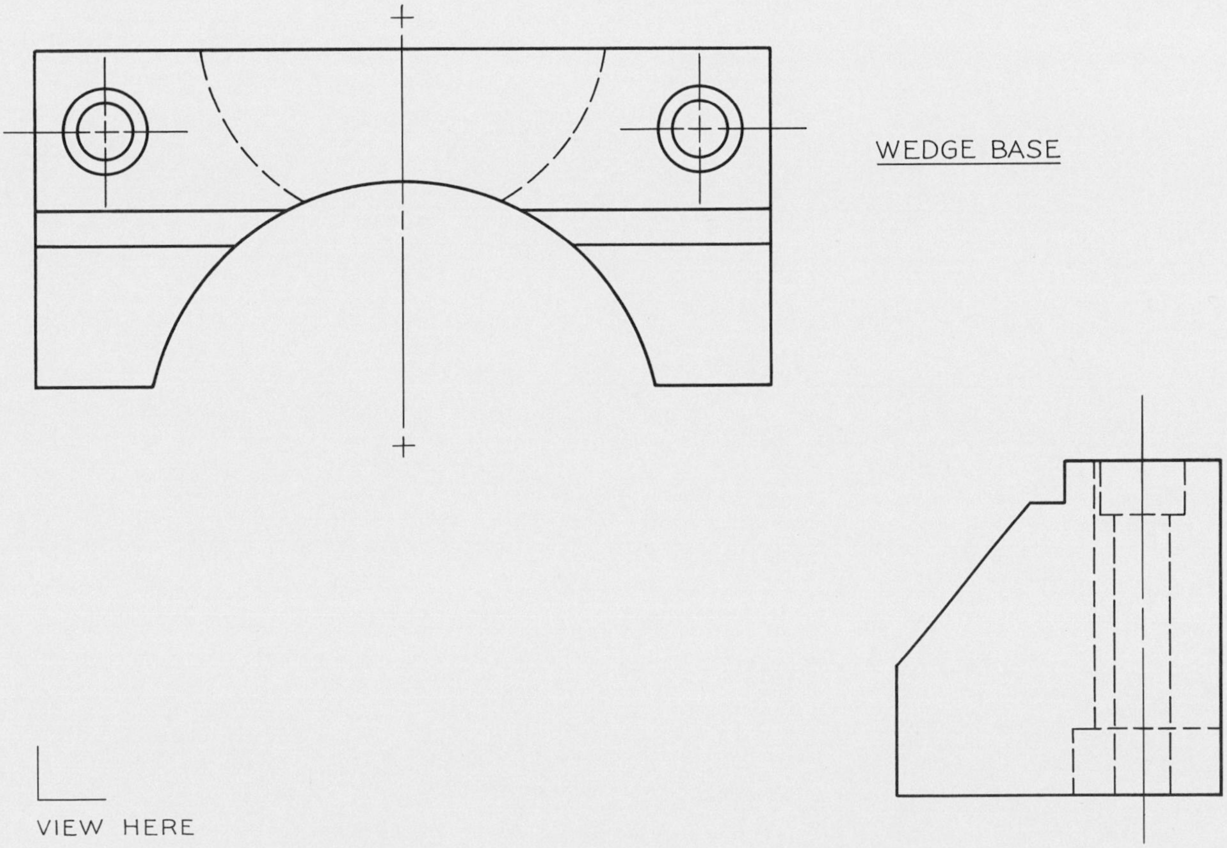
VIEW HERE

WEDGE BASE

SECTION	DRAWN BY:		SHEET 33

(1)

VIEW HERE

ANCHOR BLOCK

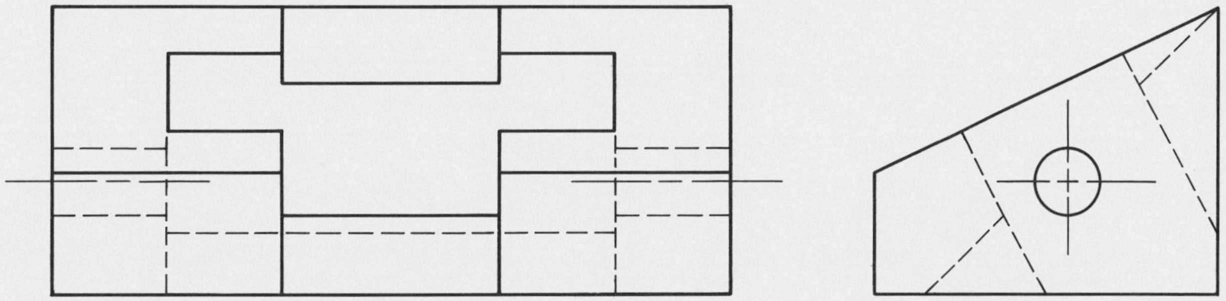

(2)

VIEW HERE

CONTROL BASE

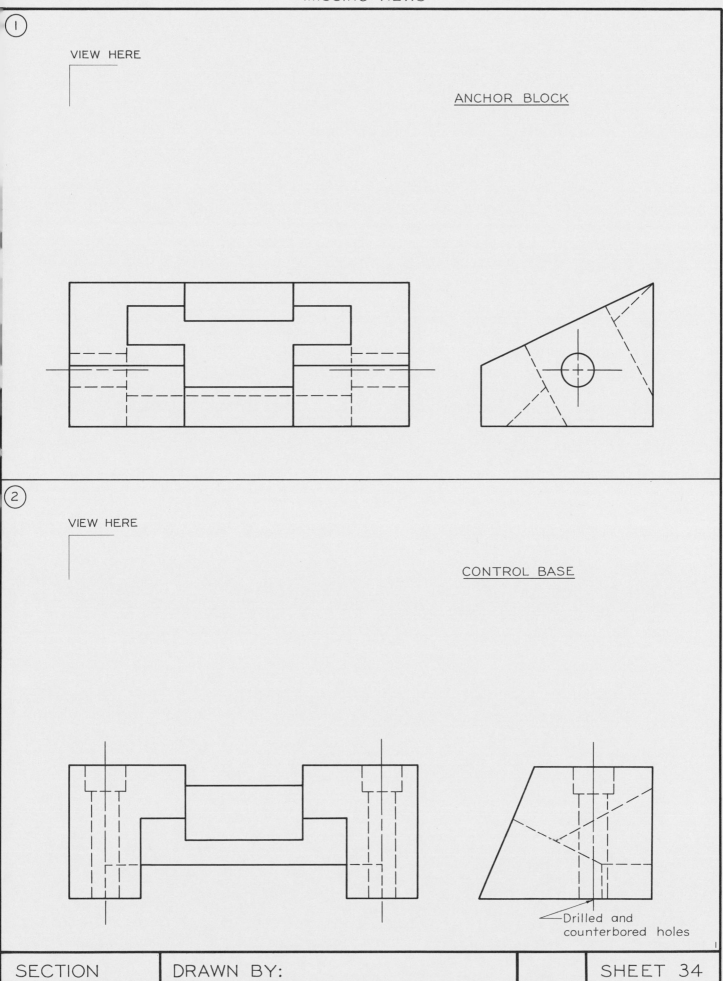

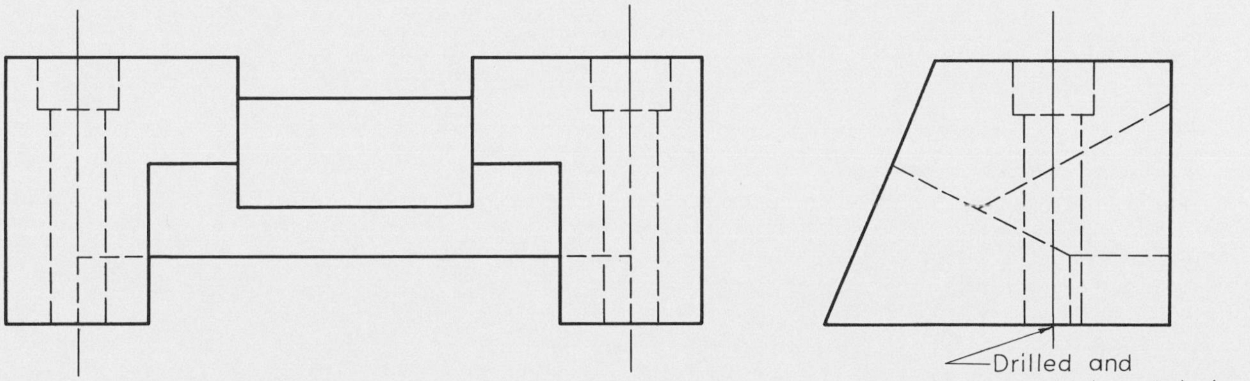
Drilled and
counterbored holes

SECTION	DRAWN BY:		SHEET 34

OPTIONAL VIEWS AND DESCRIBED SHAPE

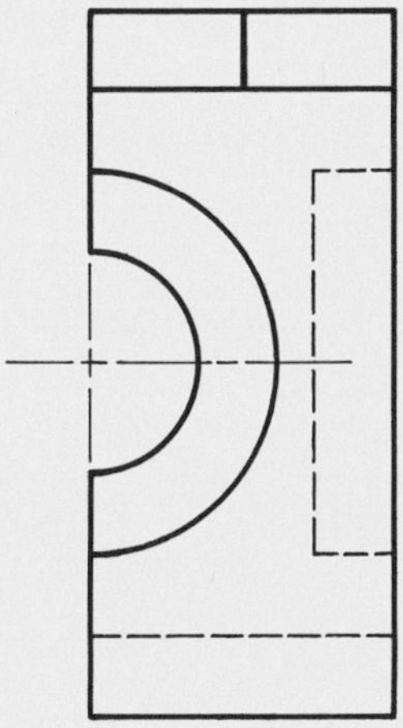

(2)

(4)

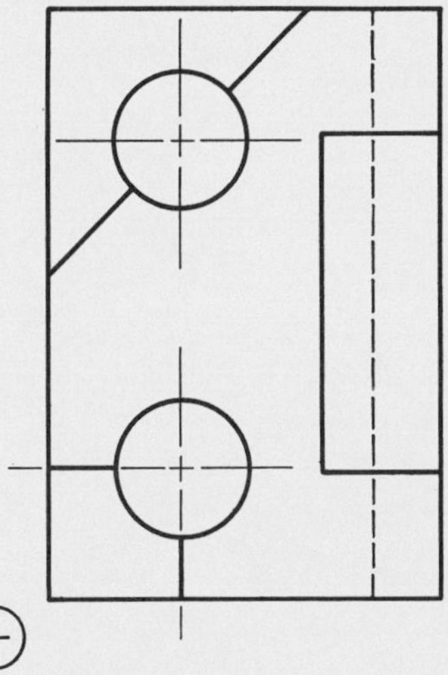

(1)

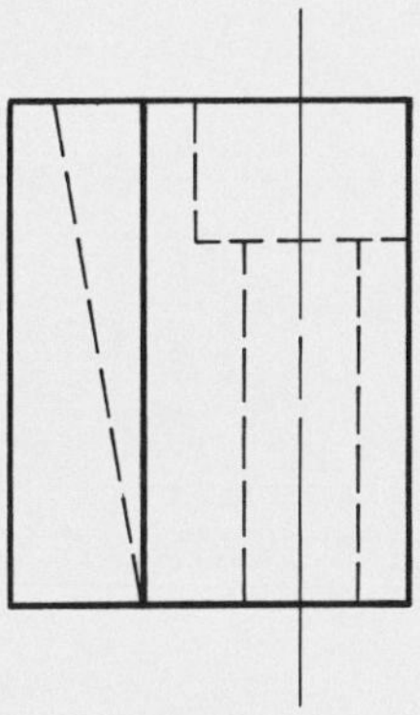

(3)

SECTION	DRAWN BY:		SHEET 36

TRAVERSE STOP PISTON

Given: Front and right-side views.
Draw top view.

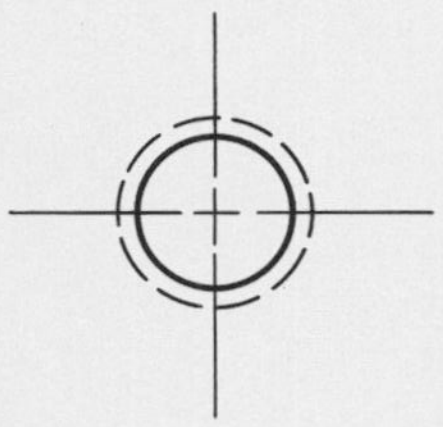

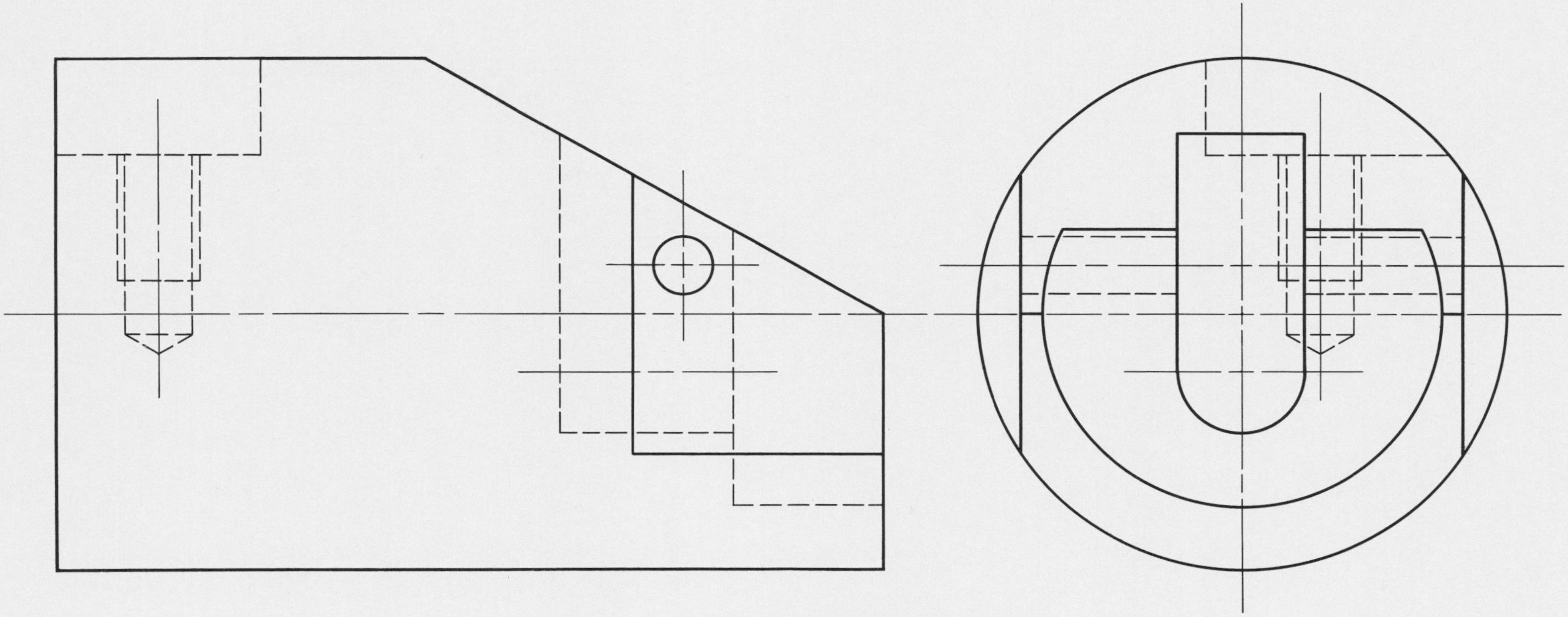

BOTTOM OF DRAWING

SECTION	DRAWN BY:		SHEET 38

CONTROL ARM

Draw right-side view.

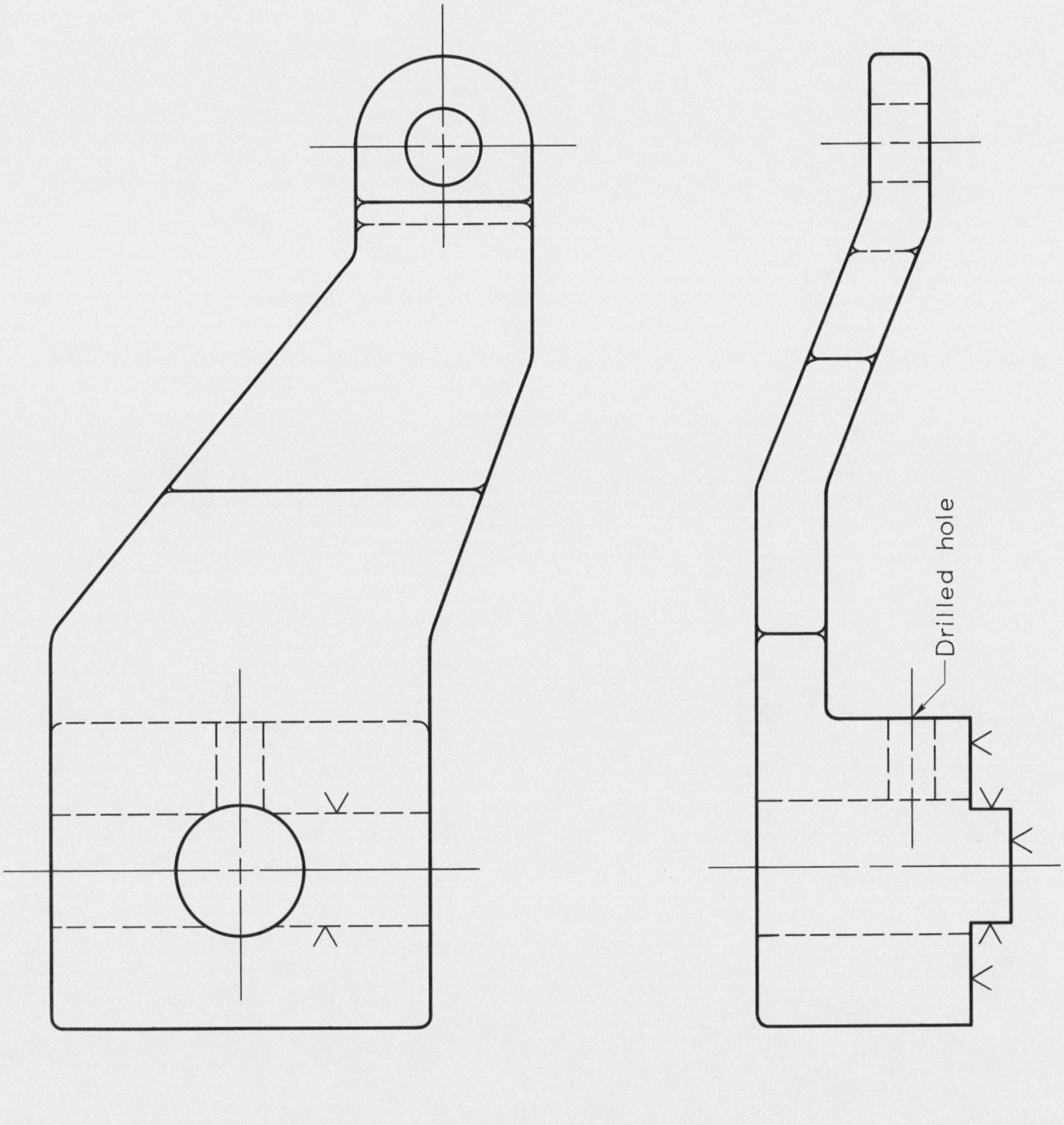
Drilled hole

BOTTOM OF DRAWING

SECTION	DRAWN BY:		SHEET 39

(1)

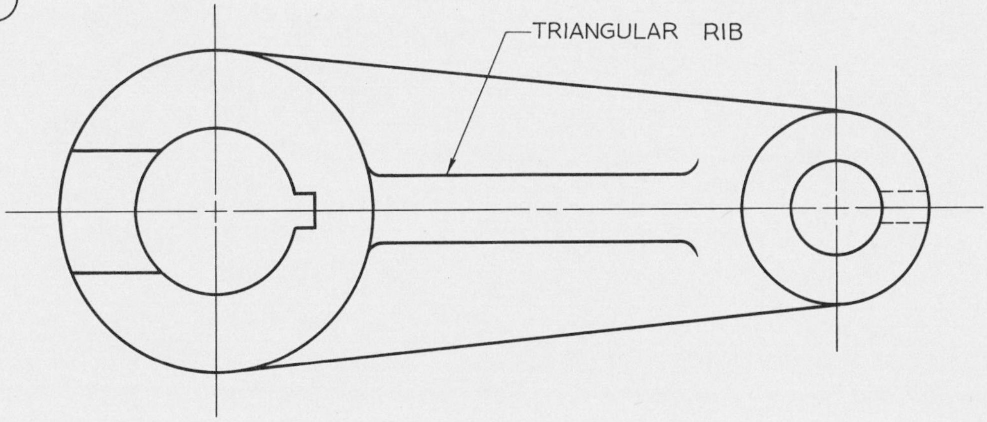
TRIANGULAR RIB

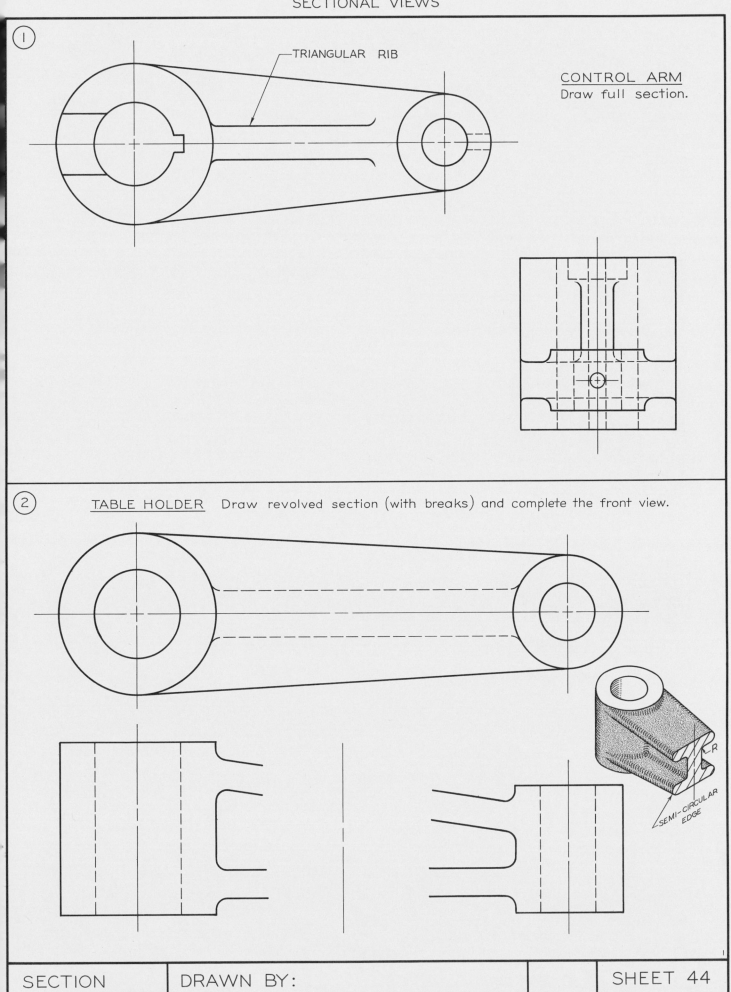

CONTROL ARM
Draw full section.

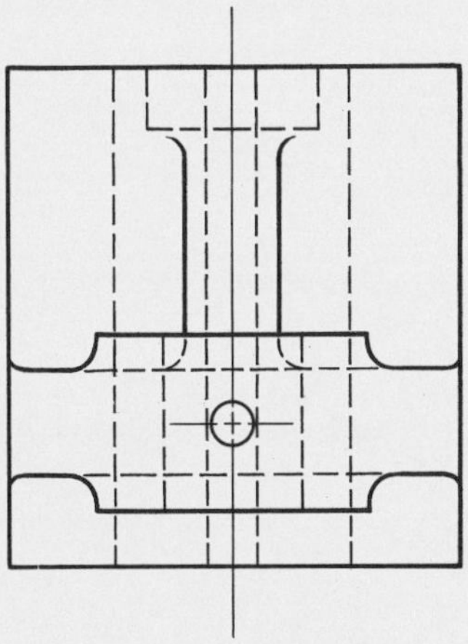

(2) TABLE HOLDER Draw revolved section (with breaks) and complete the front view.

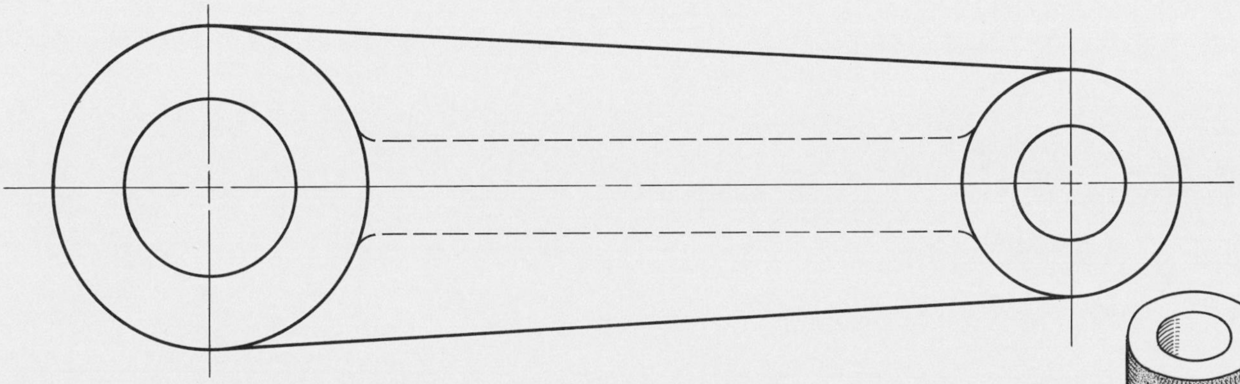

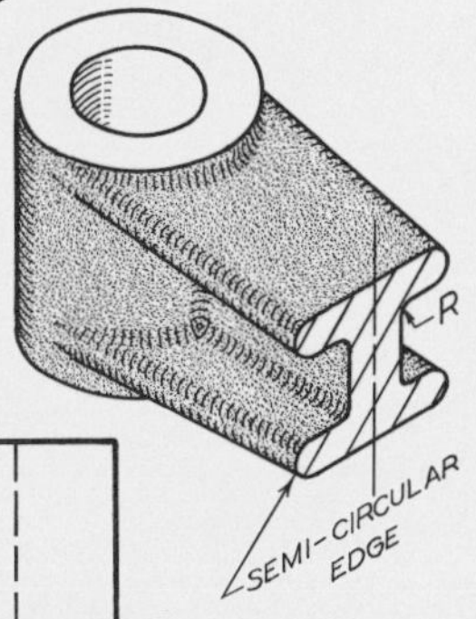
R
SEMI-CIRCULAR EDGE

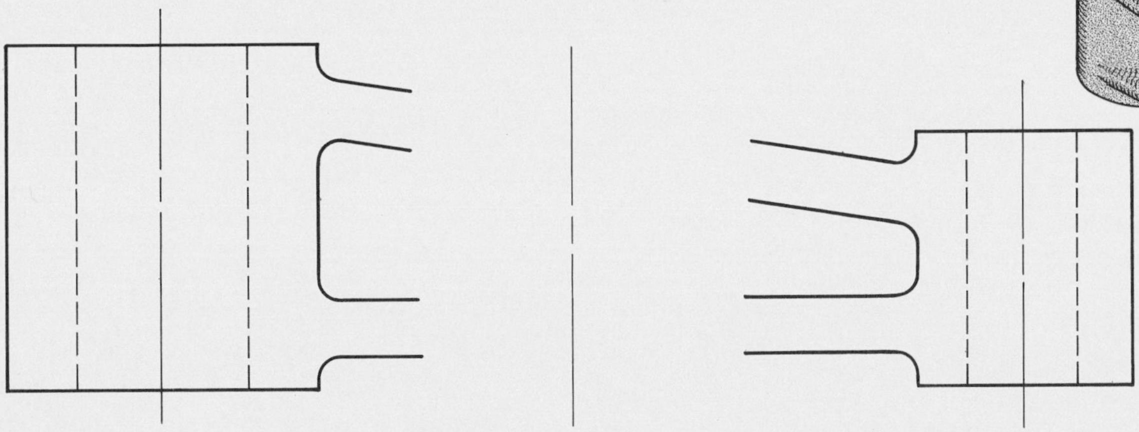

SECTION	DRAWN BY:		SHEET 44

① DRAW REVOLVED SECTIONS

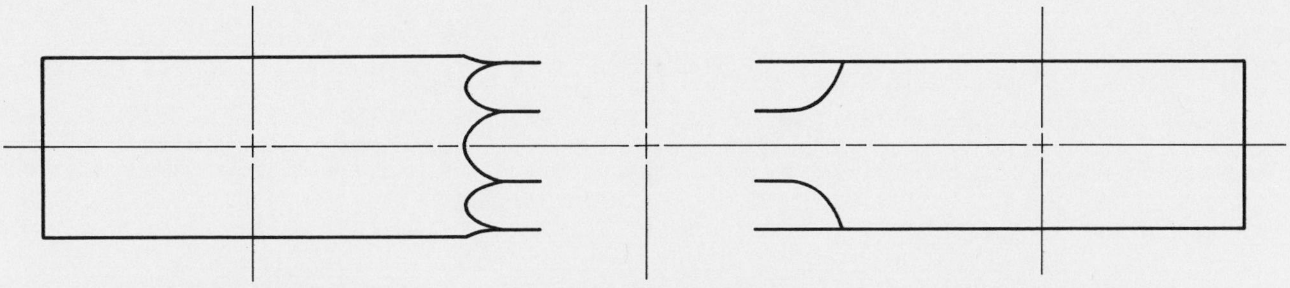
ROUND SECTION
OCTAGONAL SECTION
SQUARE SECTION

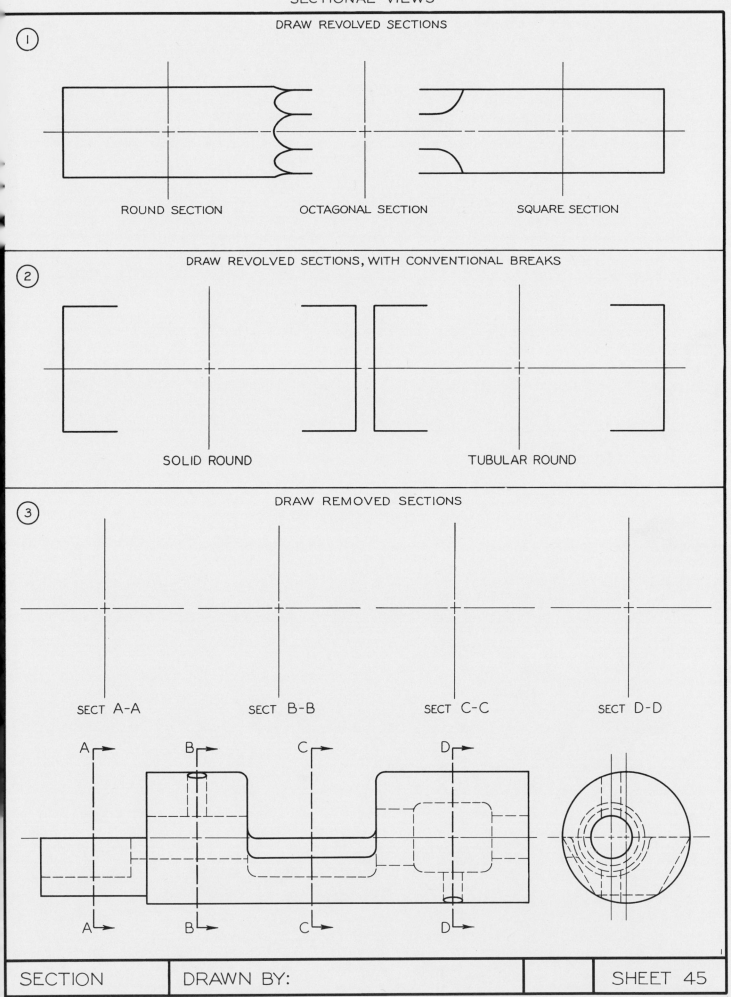

② DRAW REVOLVED SECTIONS, WITH CONVENTIONAL BREAKS

SOLID ROUND
TUBULAR ROUND

③ DRAW REMOVED SECTIONS

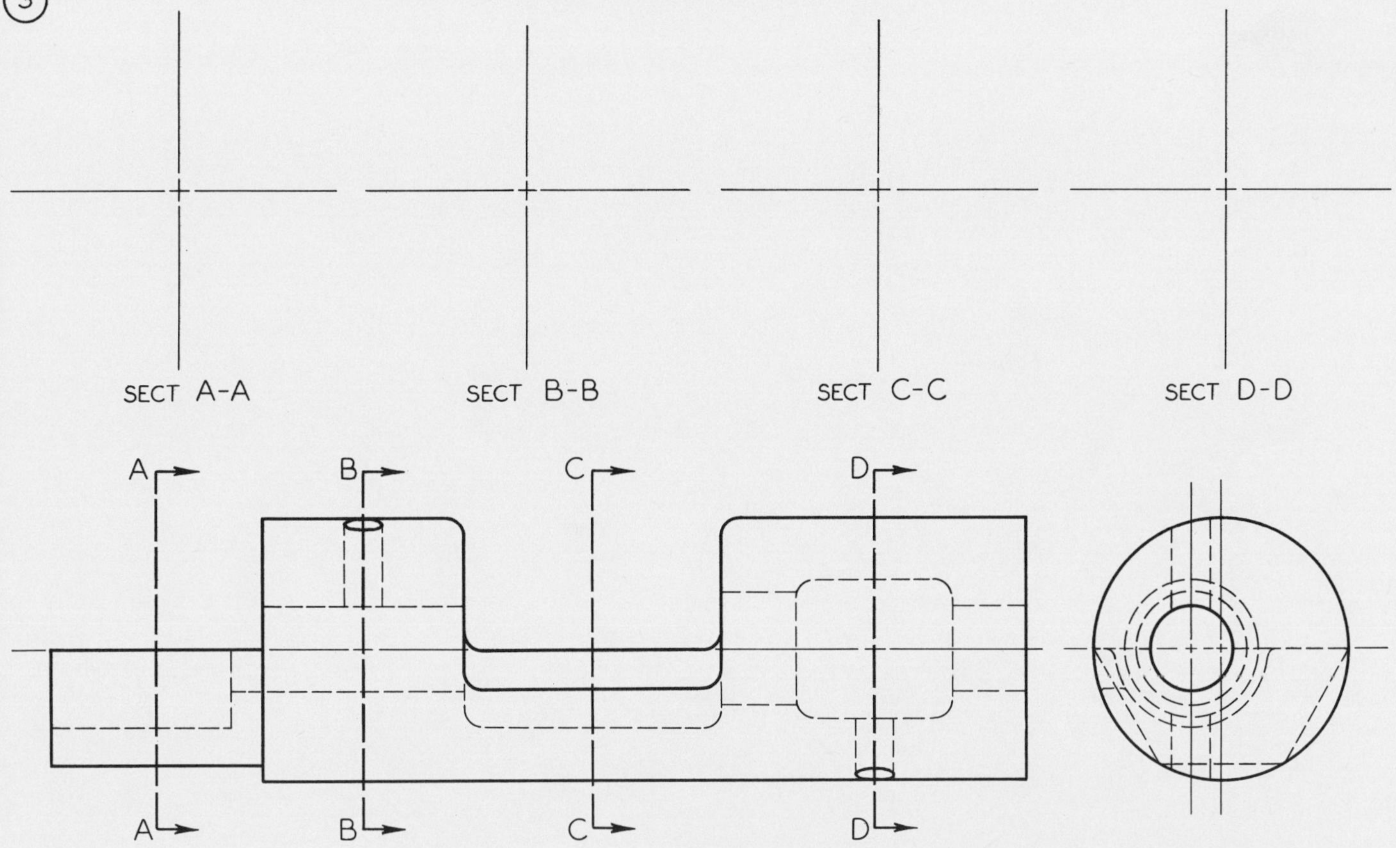
SECT A-A
SECT B-B
SECT C-C
SECT D-D
A
B
C
D
A
B
C
D

SECTION	DRAWN BY:		SHEET 45

TOOL FRAME

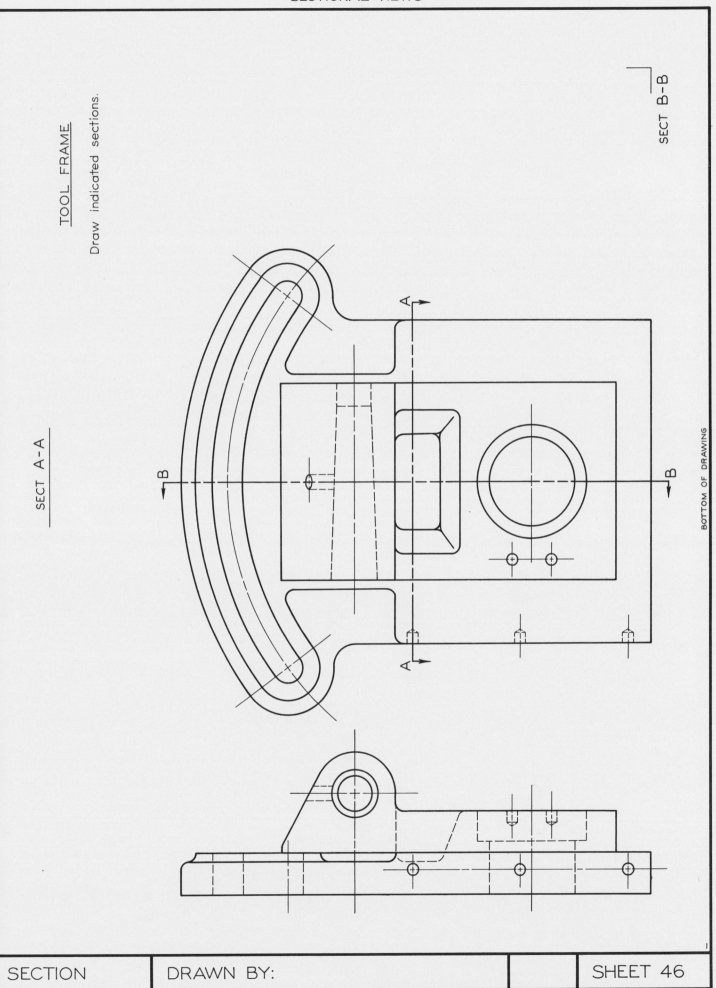

Draw indicated sections.

SECT A-A

SECT B-B

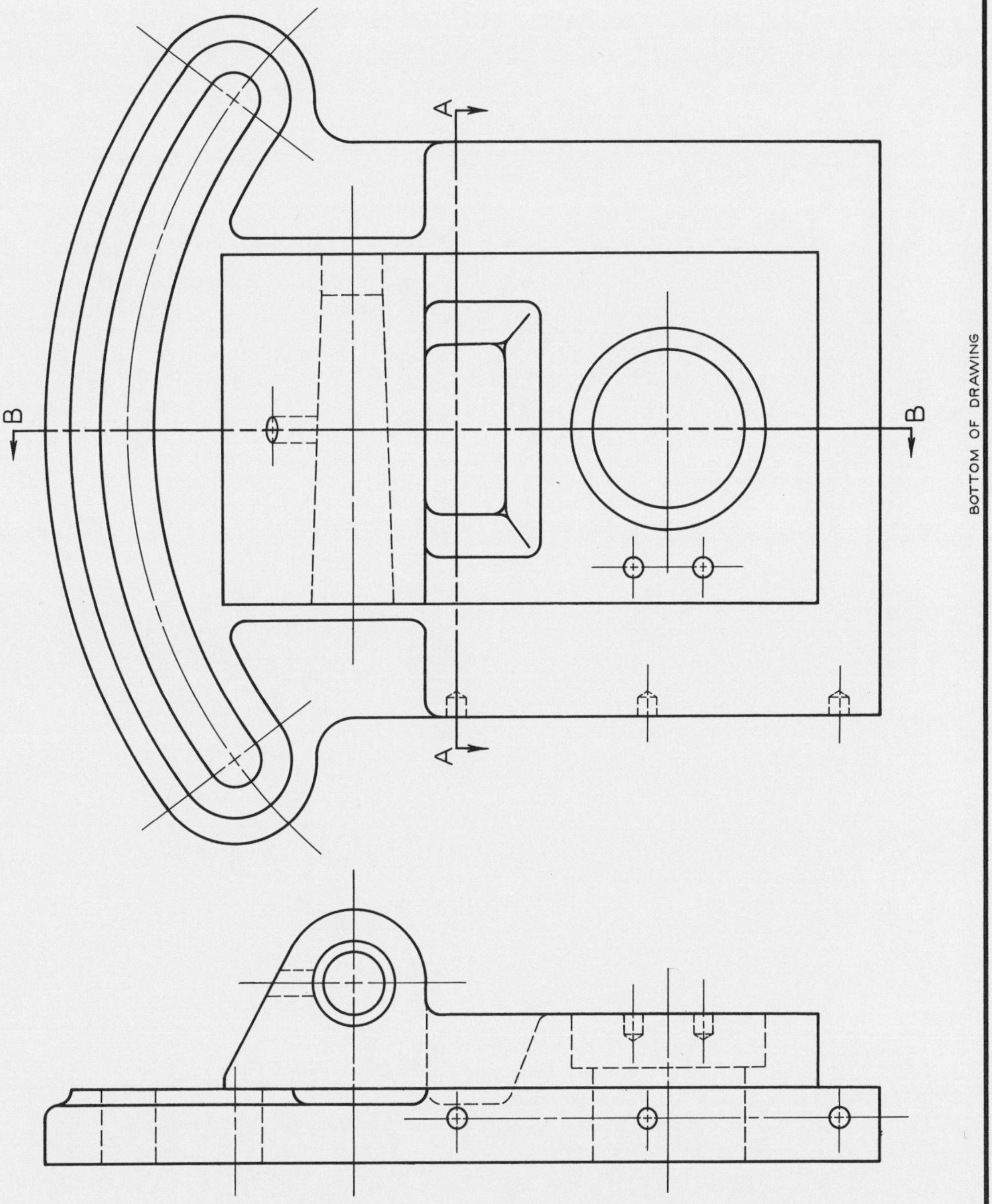
A
A
B
B
BOTTOM OF DRAWING

SECTION	DRAWN BY:		SHEET 46

LIFT BRACKET Draw indicated sections.

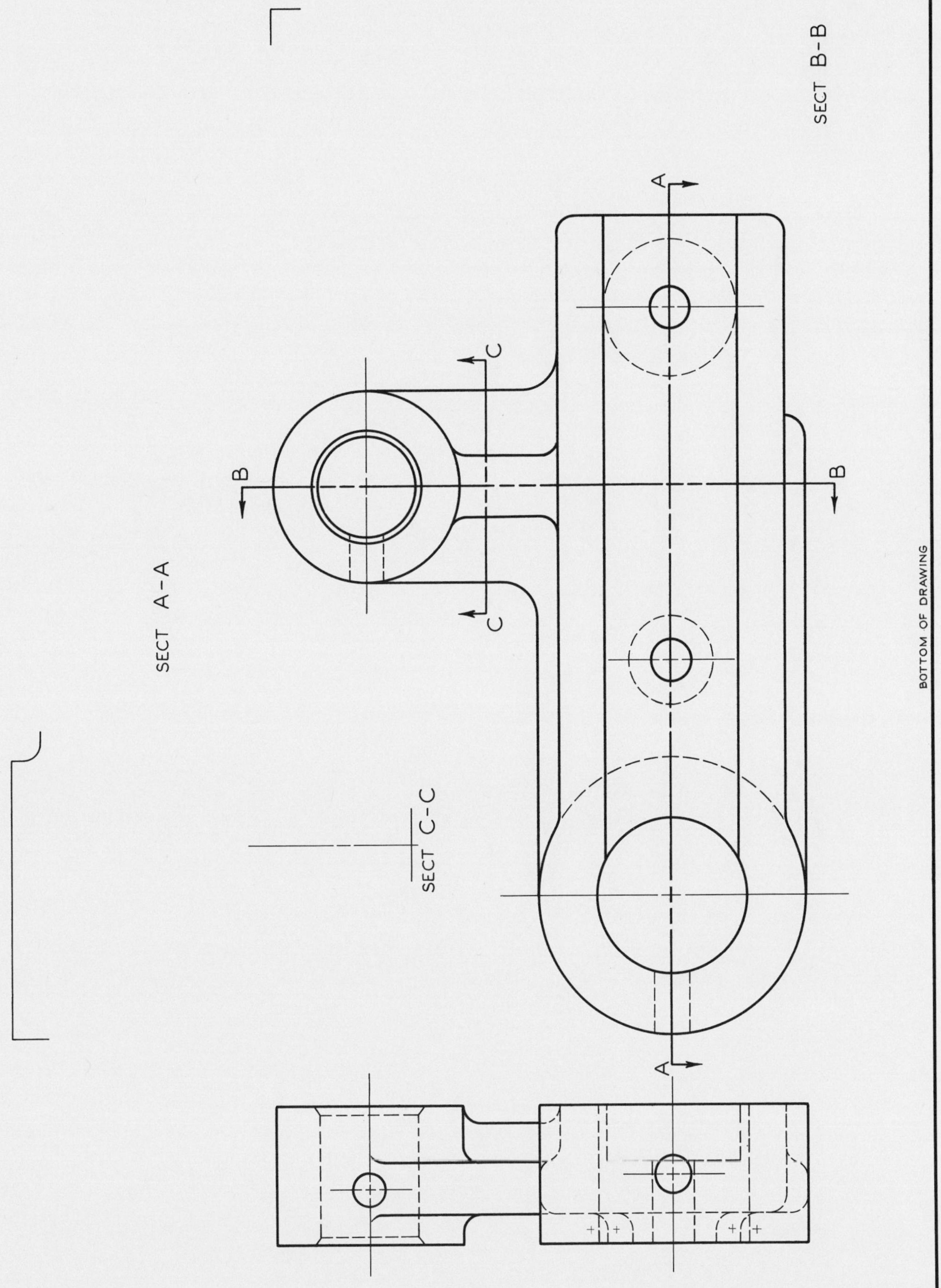
SECT B-B
SECT A-A
SECT C-C
A
A
B
B
C
C
BOTTOM OF DRAWING

SECTION	DRAWN BY:		SHEET 47

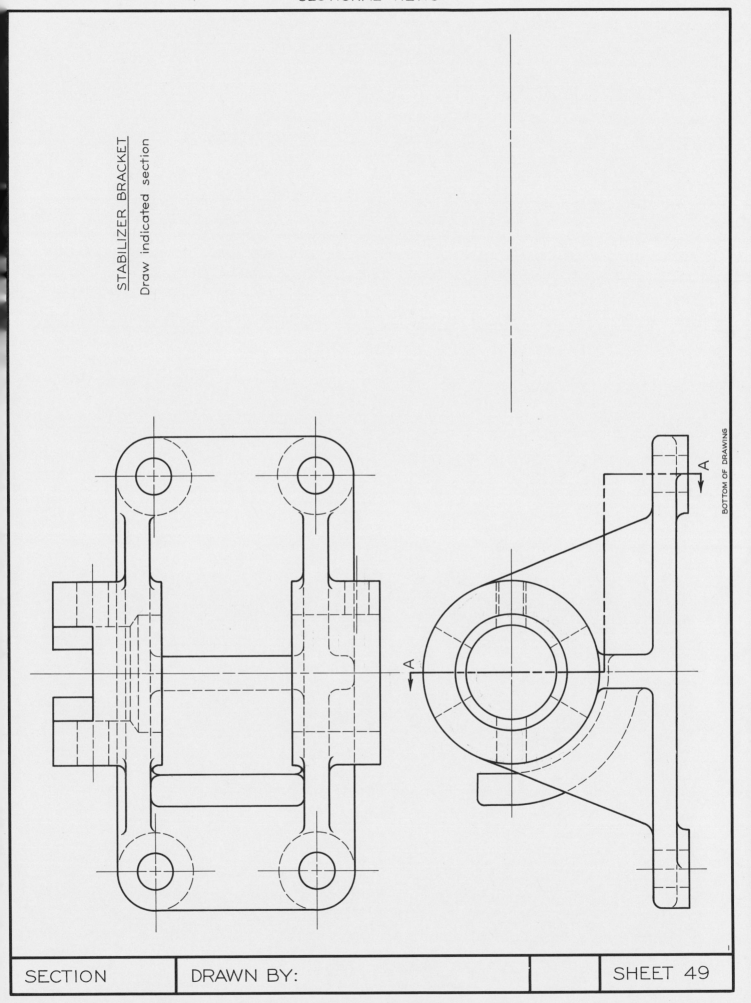

STABILIZER BRACKET

Draw indicated section

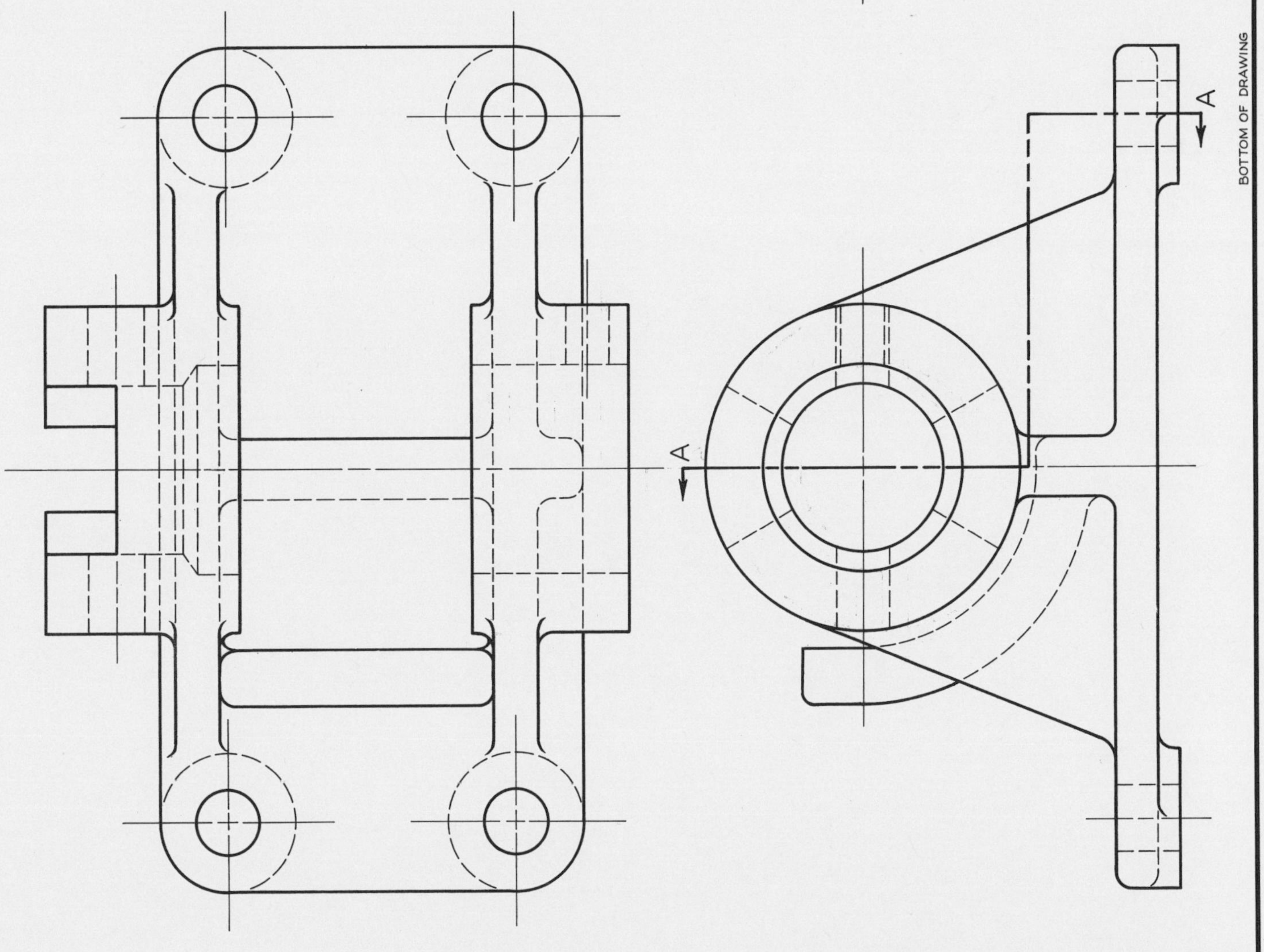
A
A
BOTTOM OF DRAWING

SECTION	DRAWN BY:		SHEET 49

SLOTTED DISC

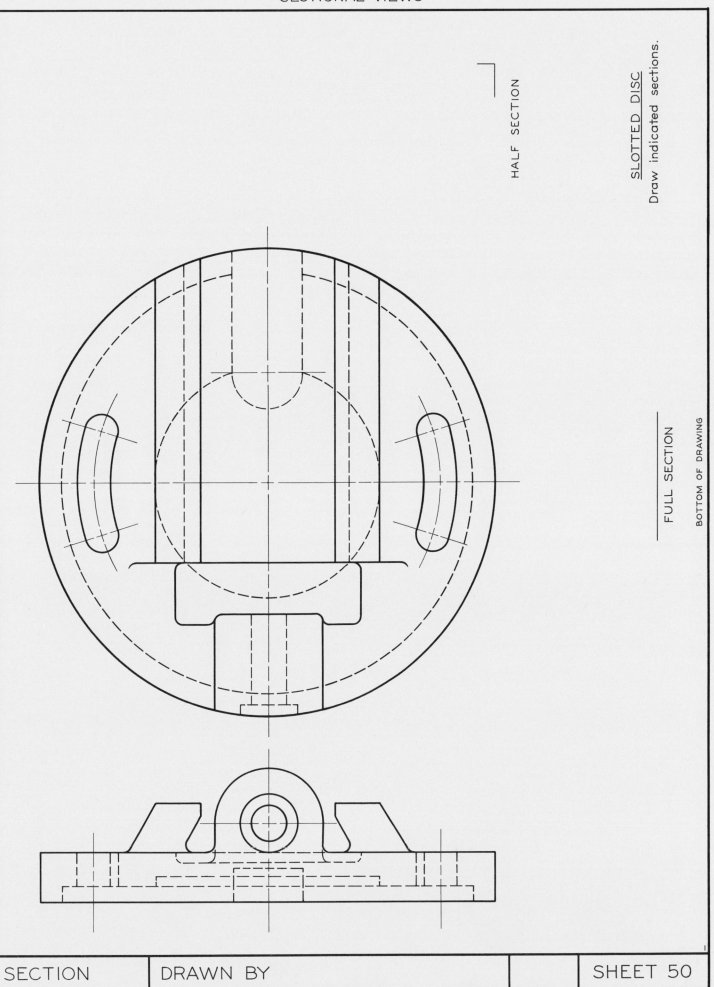

Draw indicated sections.

HALF SECTION

FULL SECTION

BOTTOM OF DRAWING

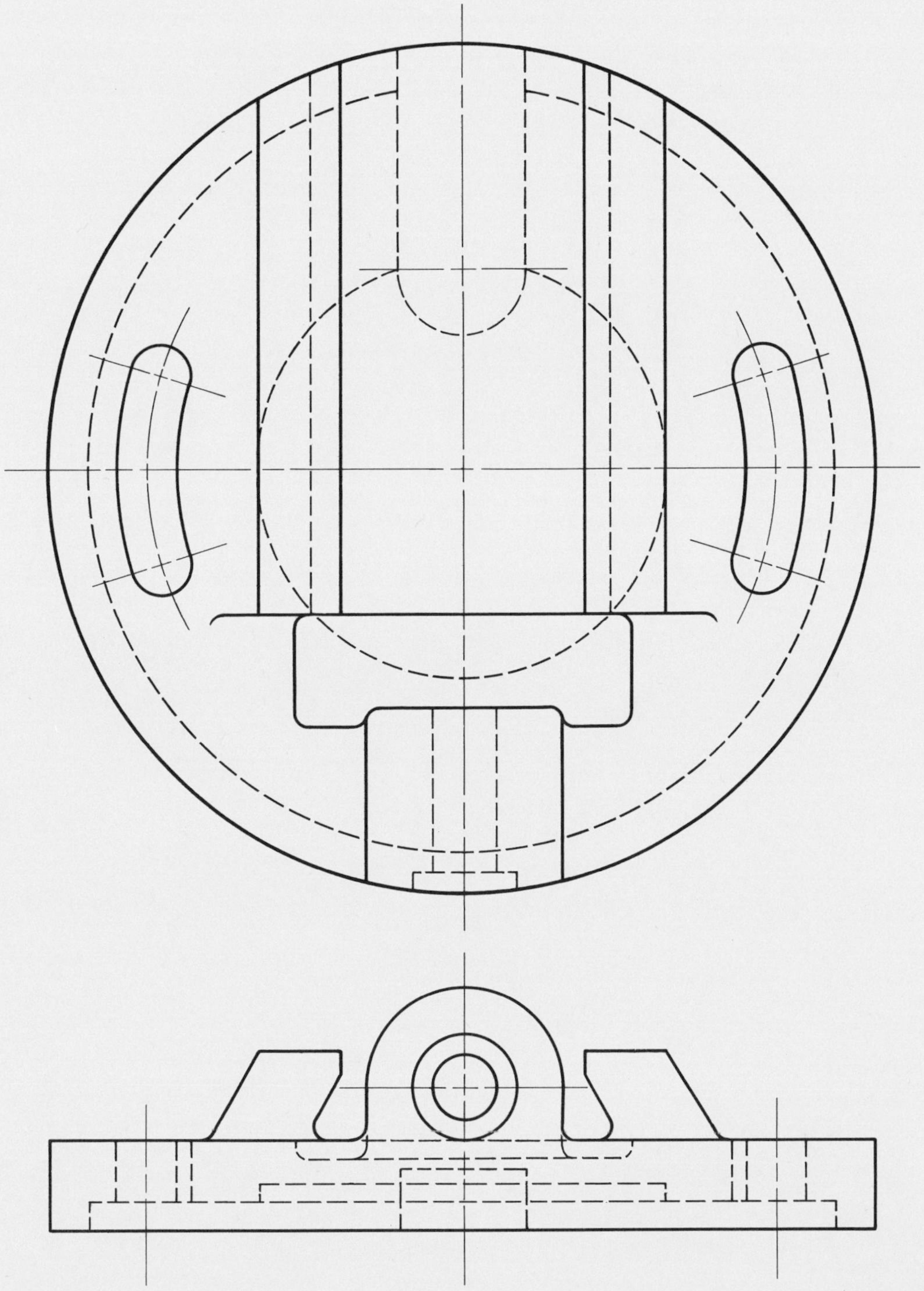

SECTION	DRAWN BY		SHEET 50

(1)

ANGLE CATCH

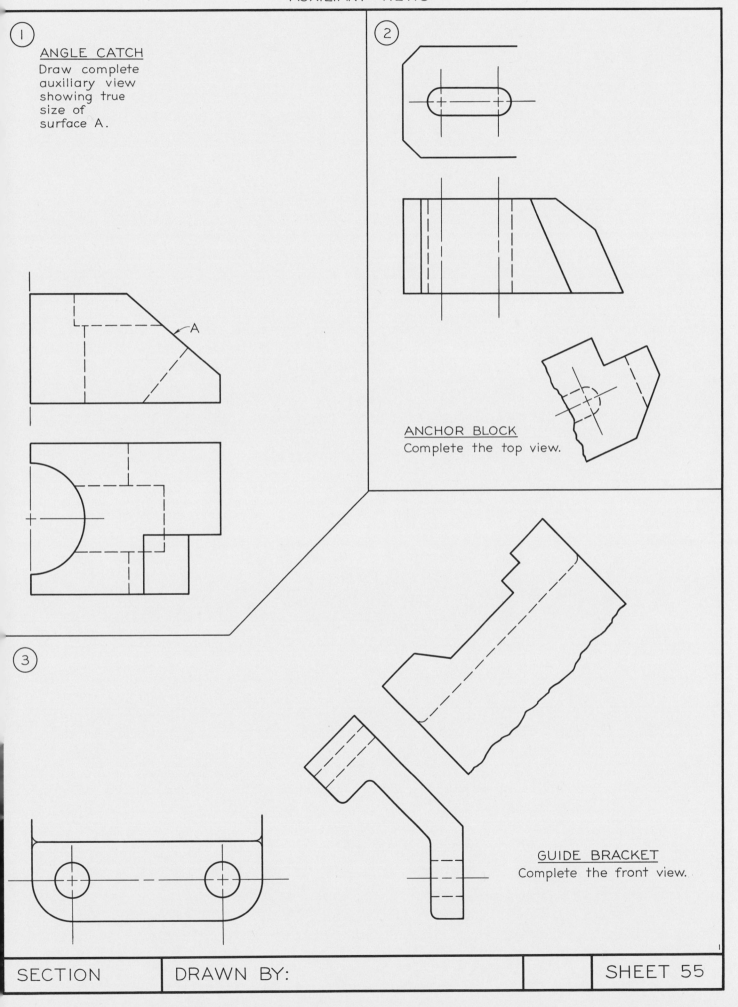

Draw complete auxiliary view showing true size of surface A.

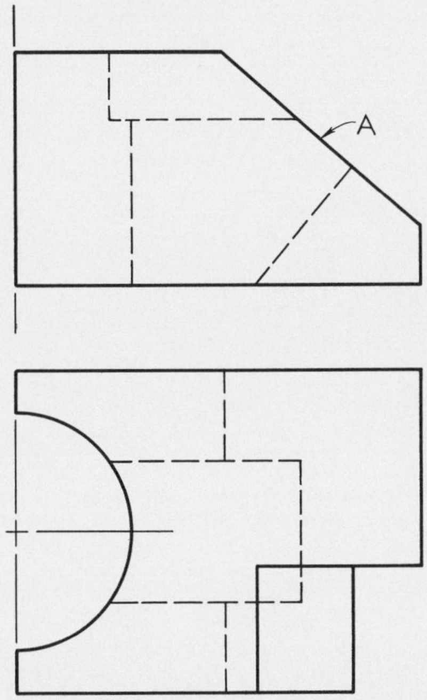
A

(2)

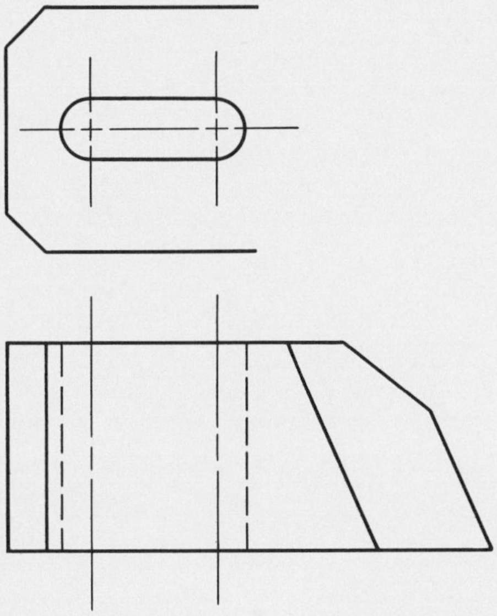

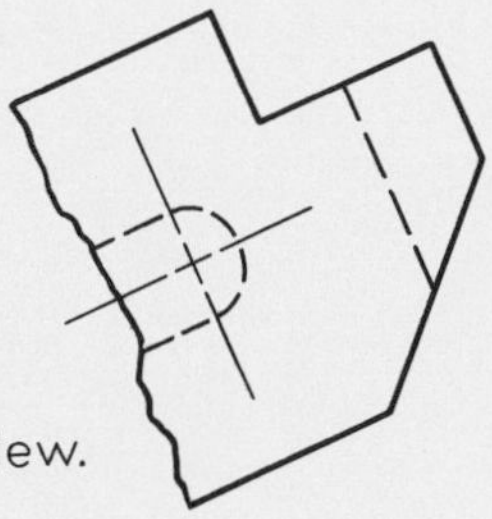

ANCHOR BLOCK

Complete the top view.

(3)

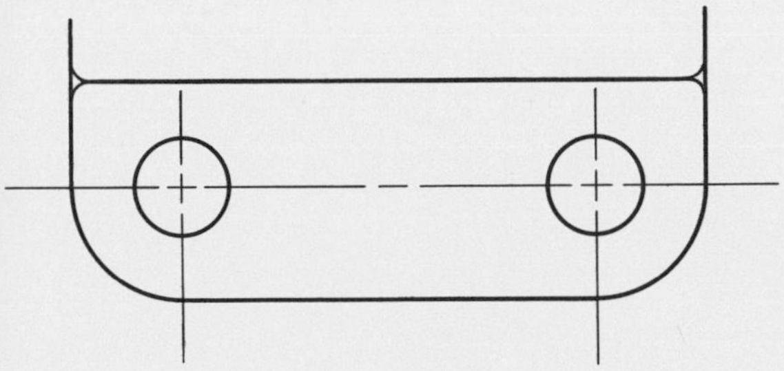

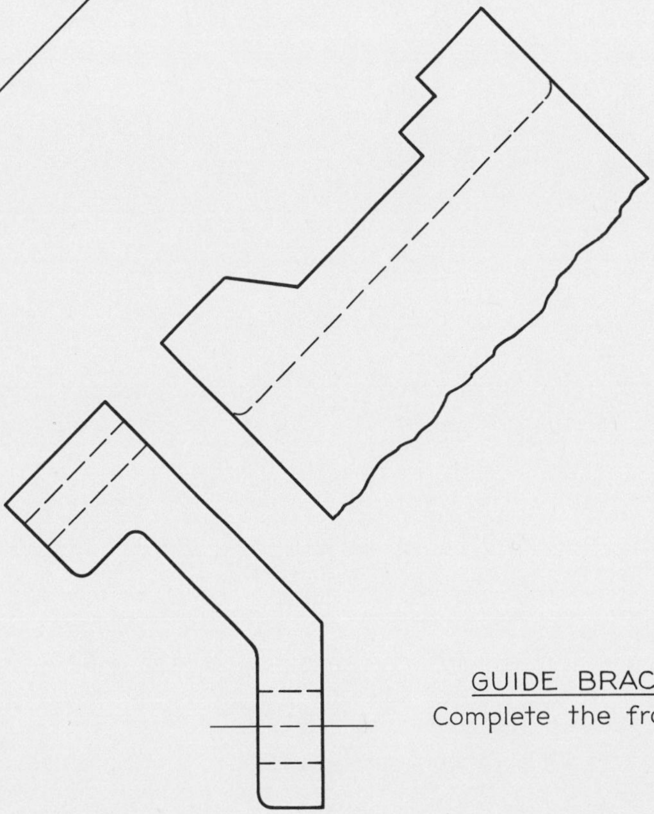

GUIDE BRACKET

Complete the front view.

SECTION	DRAWN BY:		SHEET 55

(1)

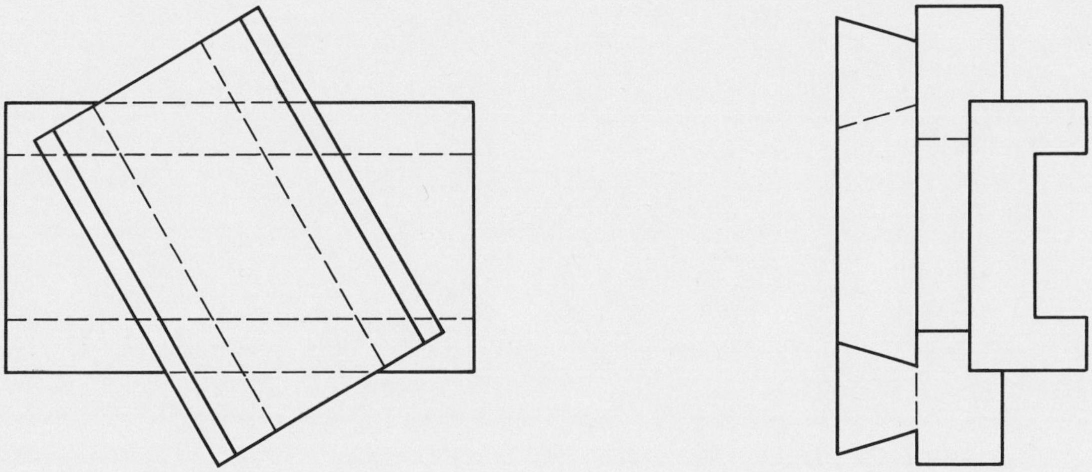

DOVETAIL SLIDE

Draw complete auxiliary view showing true angle of dovetail.

(2)

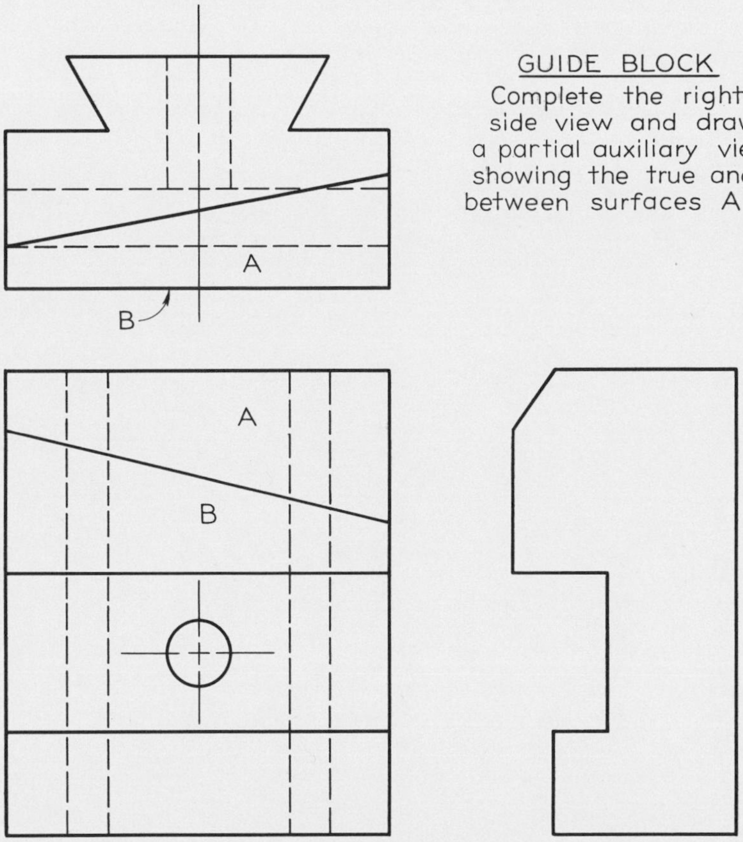
A
B
A
B

GUIDE BLOCK

Complete the right-side view and draw a partial auxiliary view showing the true angle between surfaces A & B.

(1)

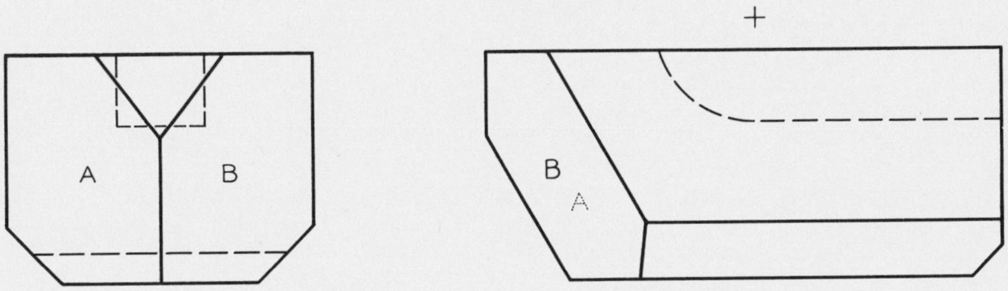
A
B
B
A

LOCATER BLOCK

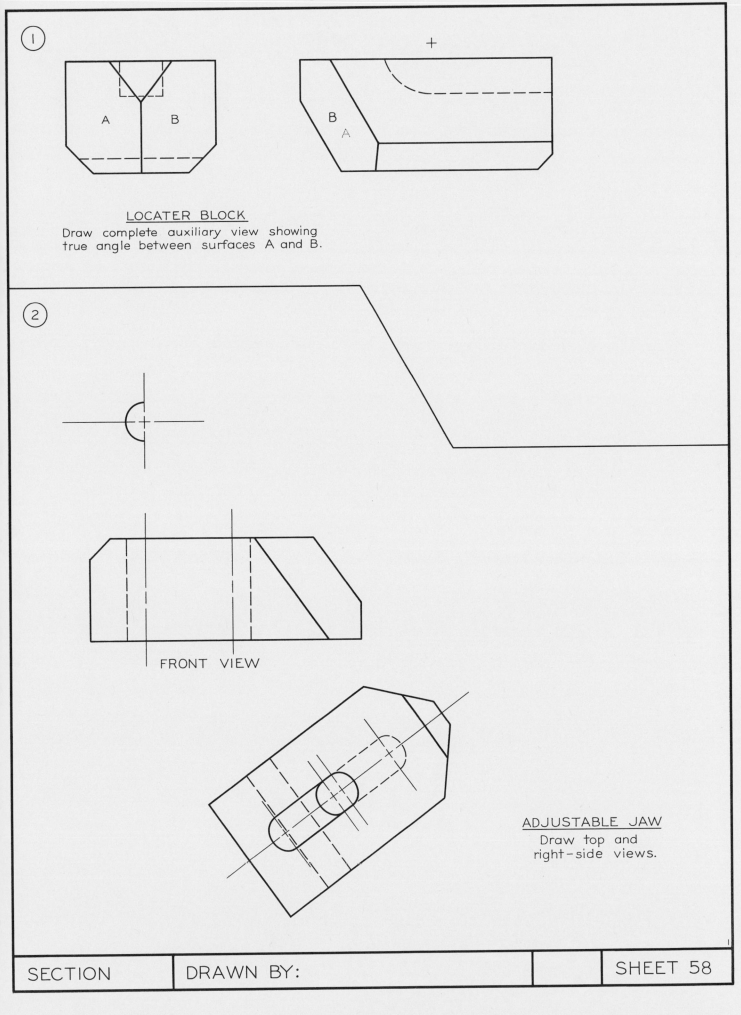

Draw complete auxiliary view showing true angle between surfaces A and B.

(2)

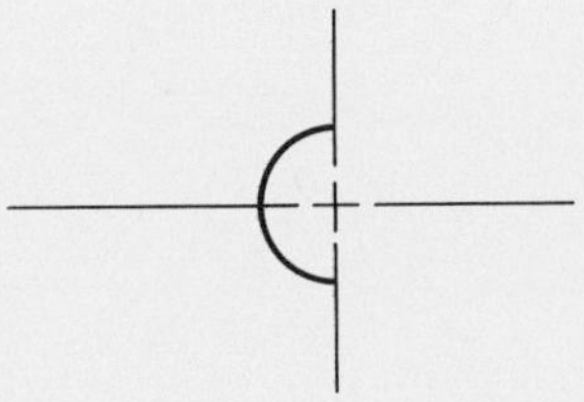

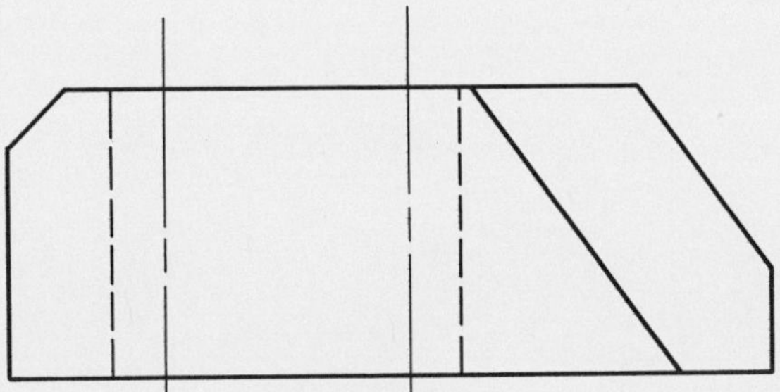

FRONT VIEW

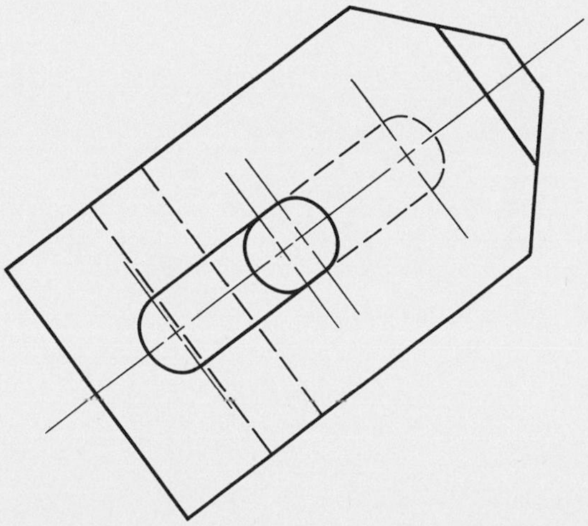

ADJUSTABLE JAW

Draw top and right-side views.

SECTION	DRAWN BY:		SHEET 58

(1)

ANGLE BRACKET

Draw secondary auxiliary view showing true size of surface A (see detail view), and complete the given views.

METRIC

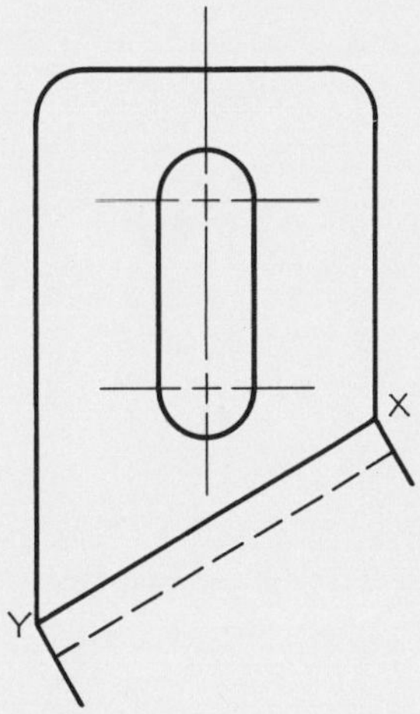

X
Y

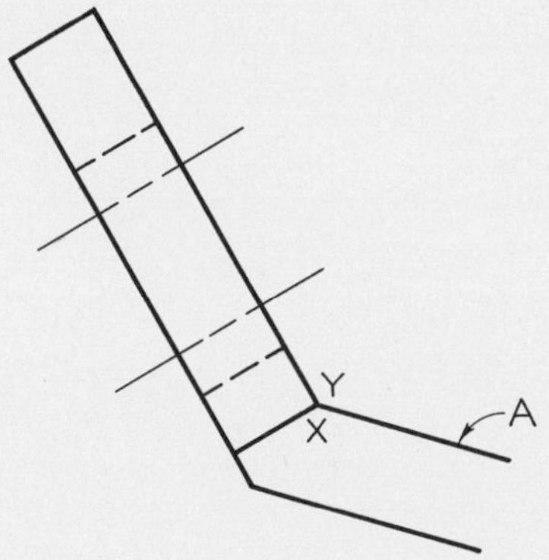

Y
X
A

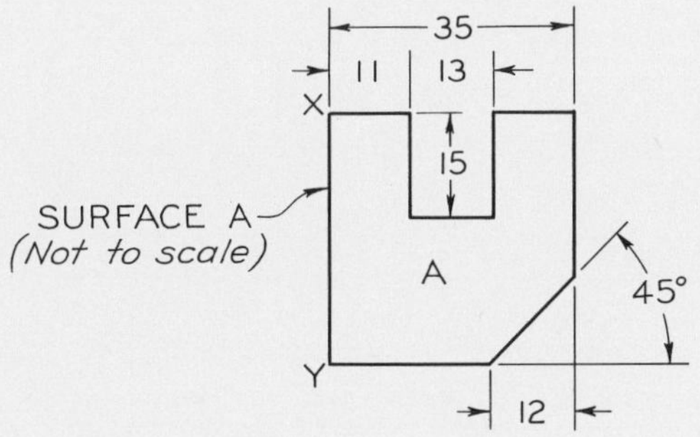

35
11
13
X
15
SURFACE A
(Not to scale)
A
45°
Y
12

(2)

CLUTCH BLOCK

From given top view draw complete auxiliary views showing true angle between surfaces A and B, and true size of surface B.

METRIC

TOP VIEW

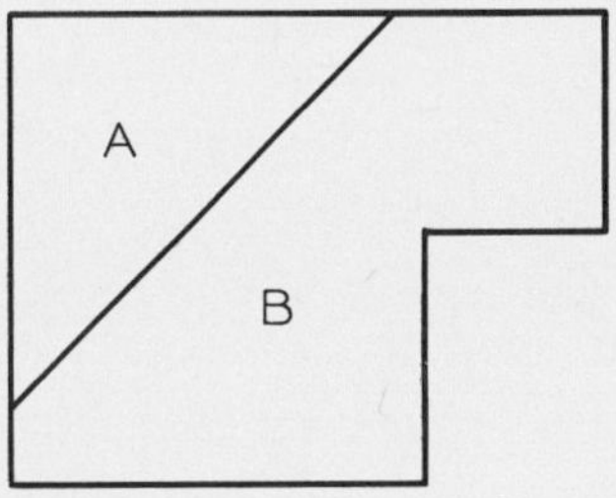

A
B

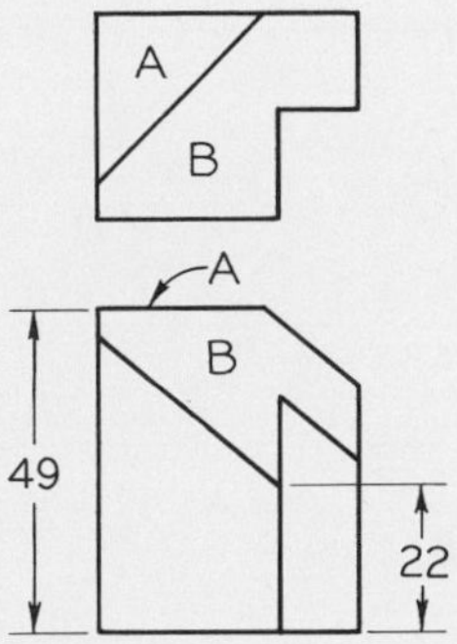

A
B
A
B
49
22

(1)

WEDGE BLOCK

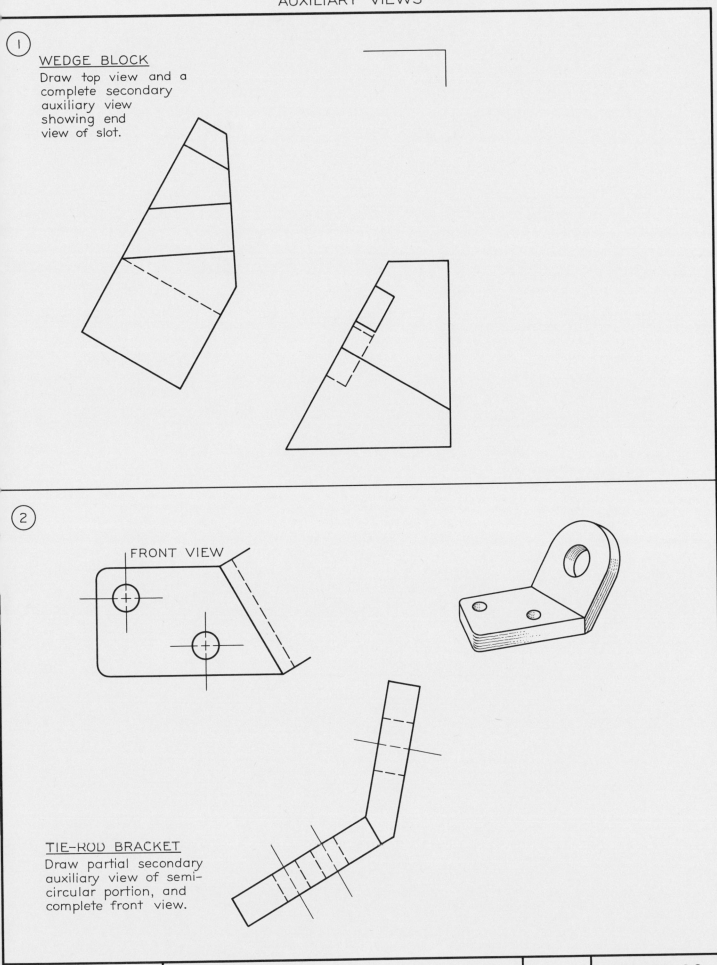

Draw top view and a complete secondary auxiliary view showing end view of slot.

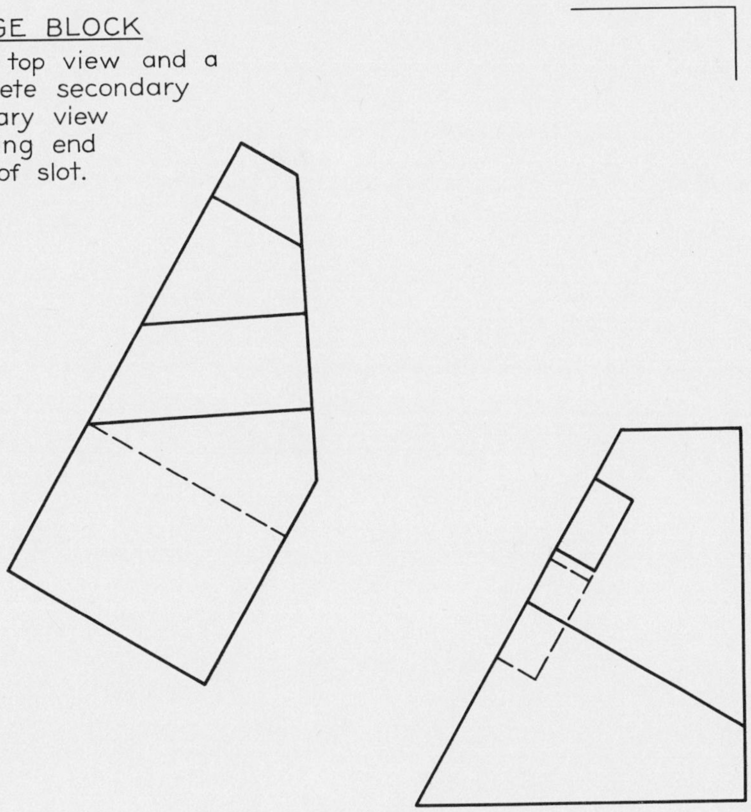

(2)

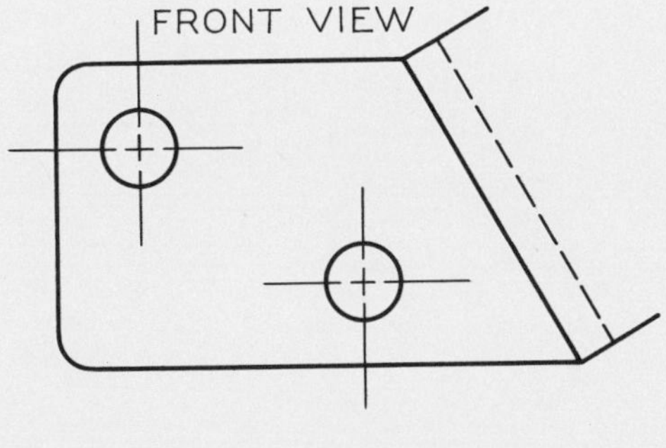
FRONT VIEW

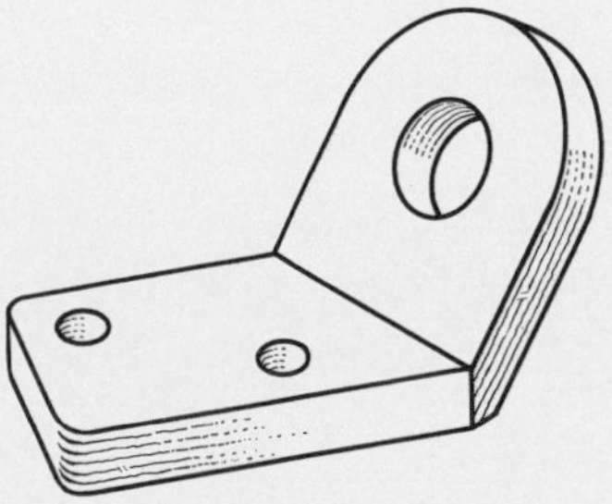

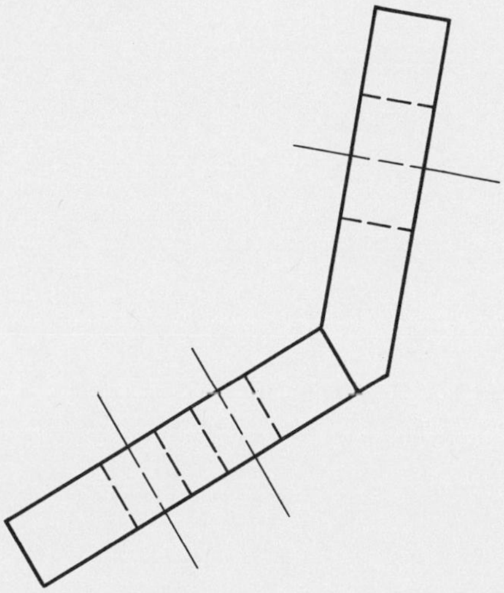

TIE-ROD BRACKET

Draw partial secondary auxiliary view of semi-circular portion, and complete front view.

SECTION	DRAWN BY:		SHEET 60

1

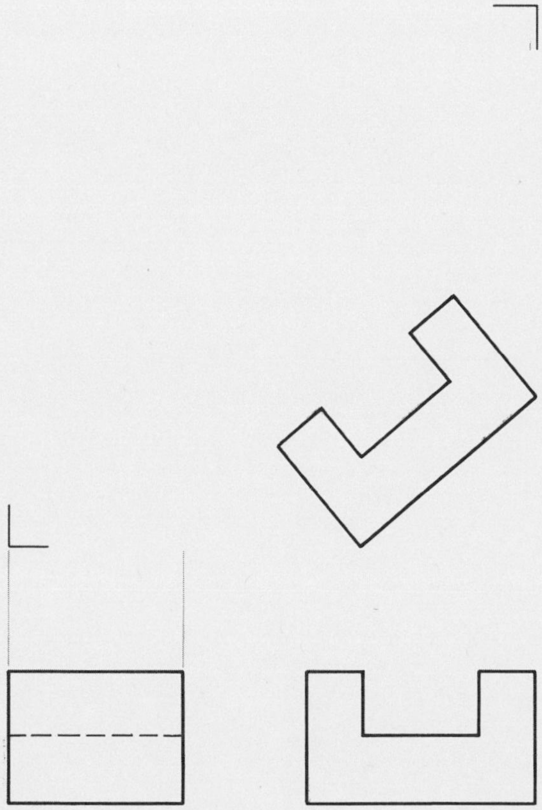

What dimension remains unchanged? ____________

2

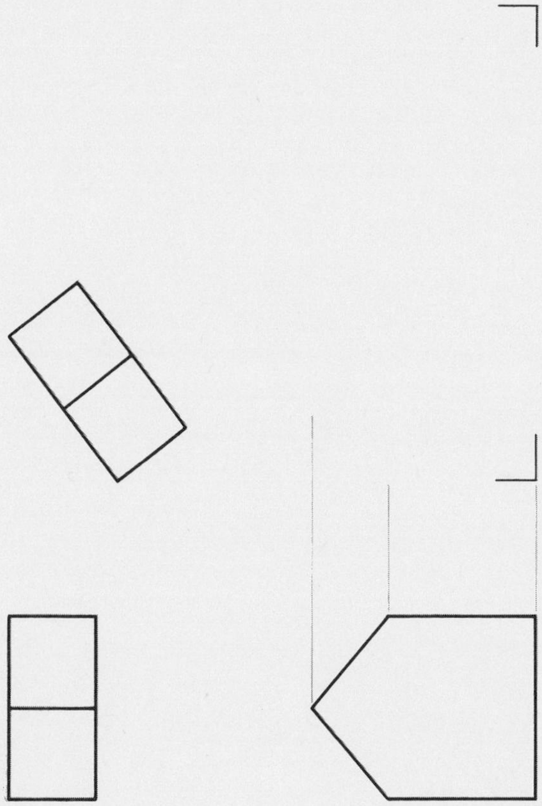

What dimension remains unchanged? ____________

3

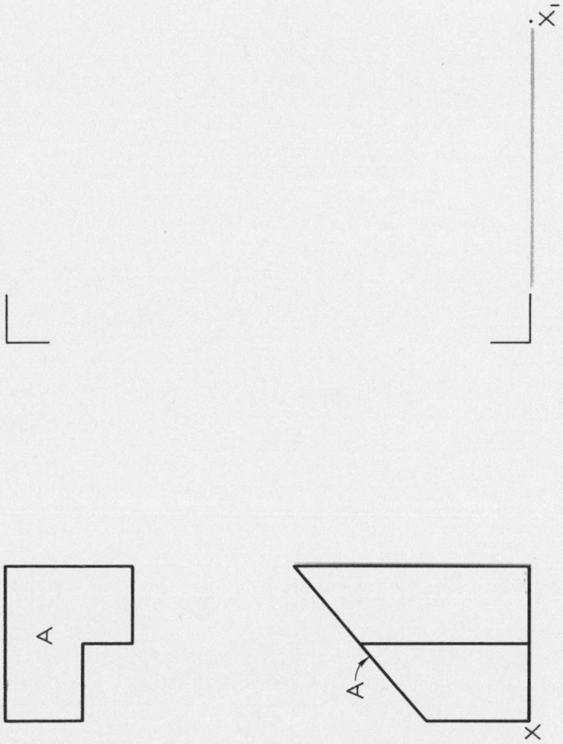
A
A
X
X1

What dimension remains unchanged? ____________

4

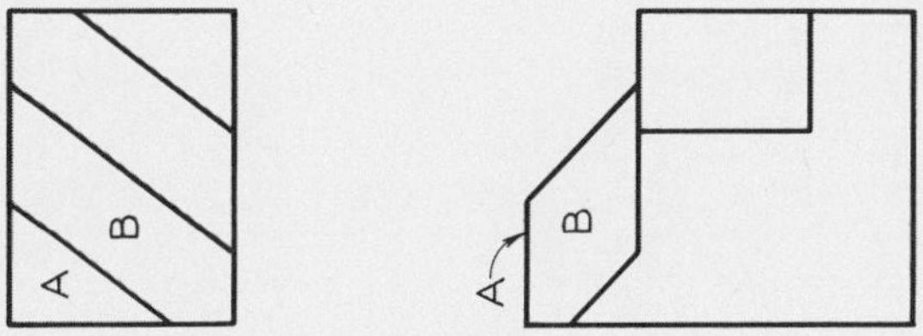
A
B
A
B

In which view does axis appear as a point? ____________

BOTTOM OF DRAWING

SECTION	DRAWN BY:		SHEET 62

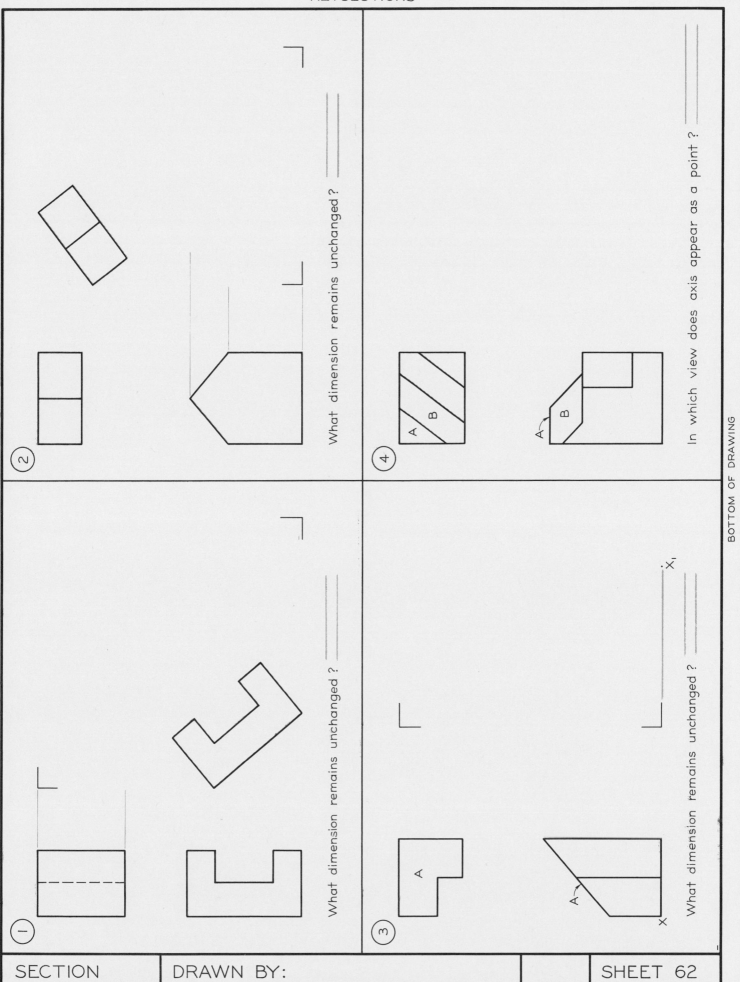

REVOLUTIONS

(1)

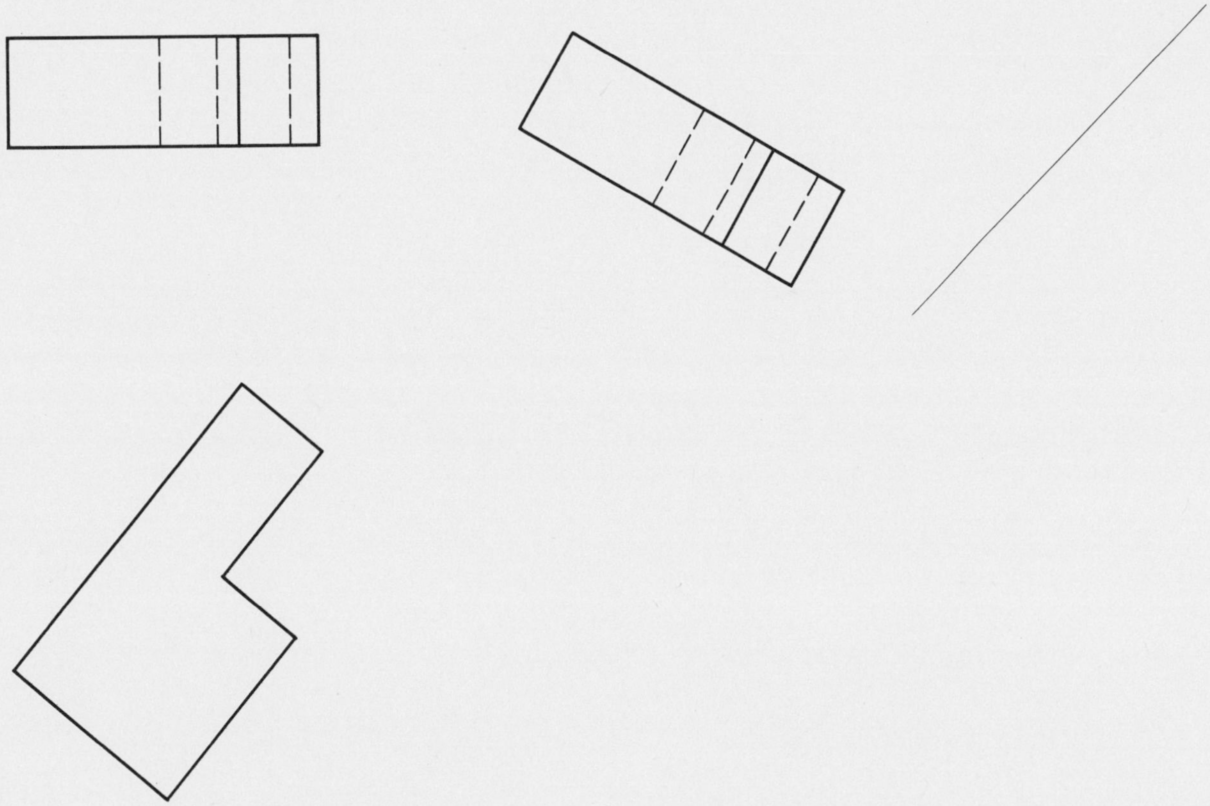

(2)

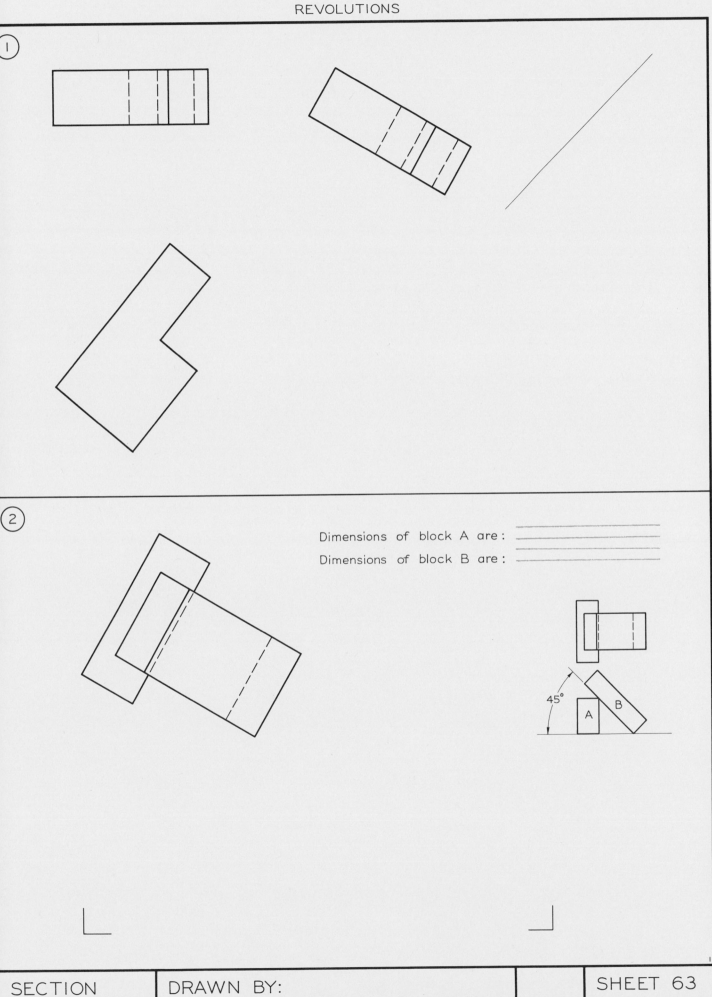

Dimensions of block A are: ______________________

Dimensions of block B are: ______________________

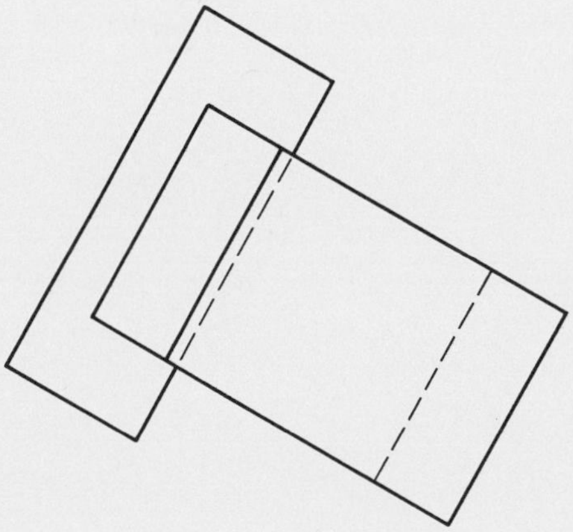

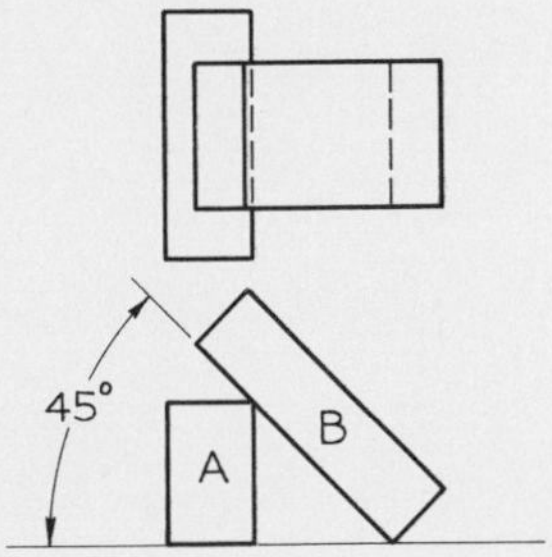
45°
A
B

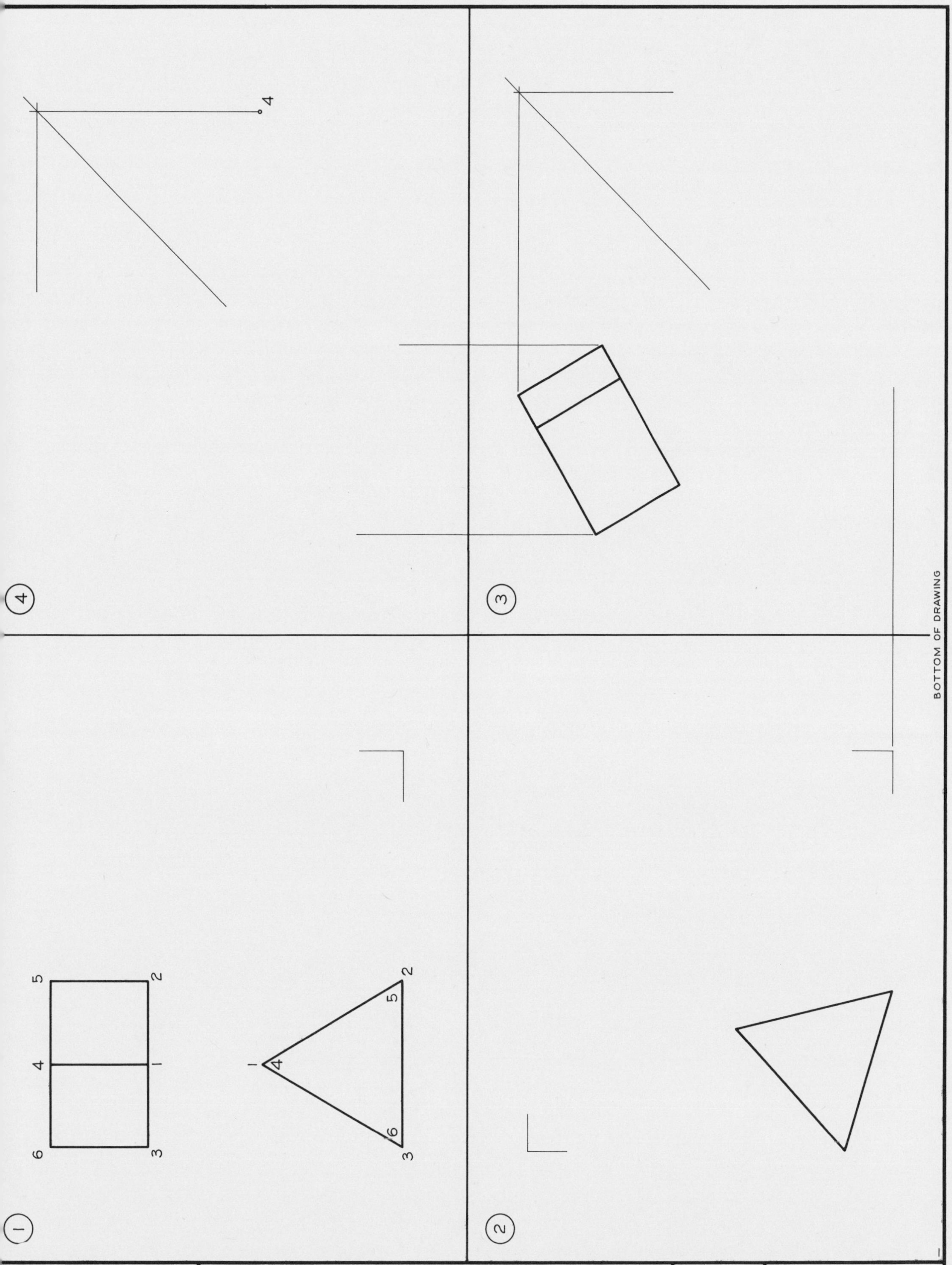
4
4
3
BOTTOM OF DRAWING
5
2
4
1
1
4
5
2
6
3
6
3
1
2

SECTION	DRAWN BY:		SHEET 64

(1)

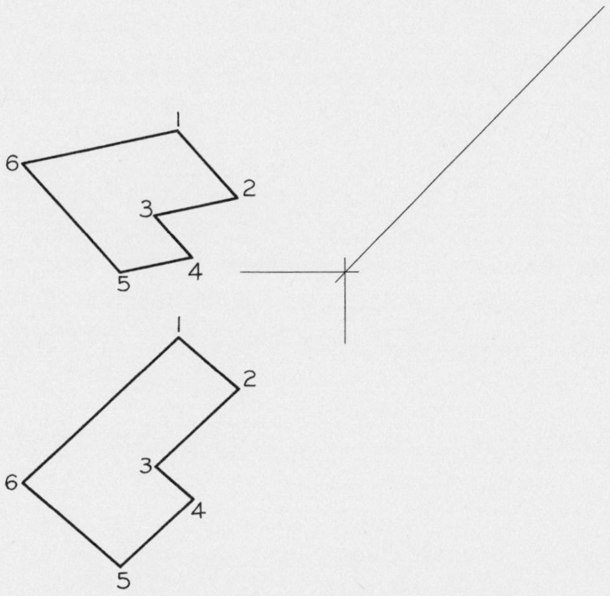
1
6
2
3
5
4
1
2
3
6
4
5

(2)

(4)

(3)

SECTION	DRAWN BY:		SHEET 65

(1)

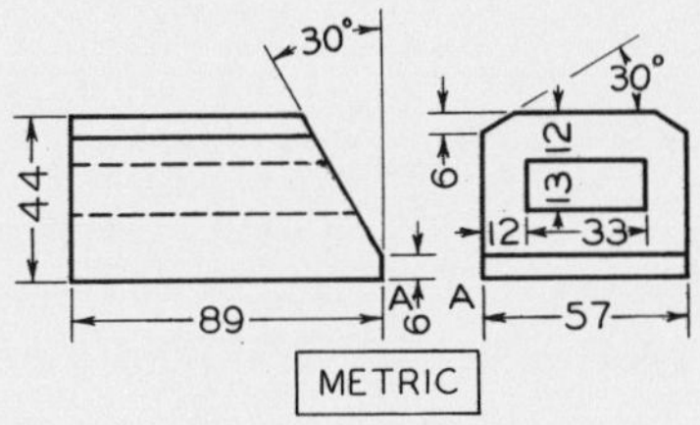
30°
30°
44
6
12
13
12
33
A
A
6
89
57
METRIC

BEVEL STOP
Draw isometric drawing.

A

(2)

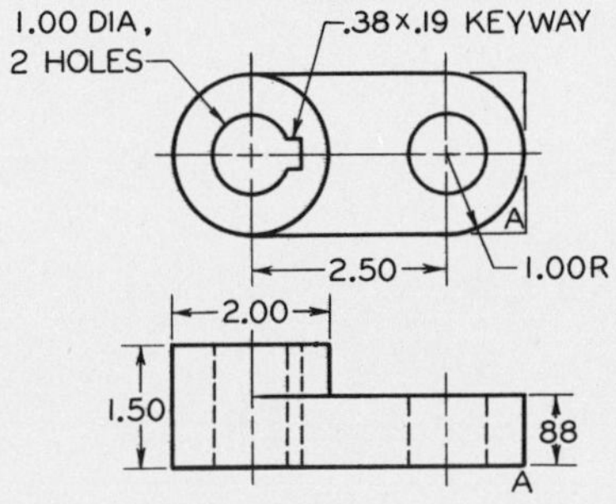
1.00 DIA,
2 HOLES
.38 X .19 KEYWAY
A
2.50
1.00R
2.00
1.50
88
A

LH LEVER
Draw isometric drawing.

A

SECTION	DRAWN BY:		SHEET 68

(1)

LOCK CLIP

Draw isometric drawing.

A

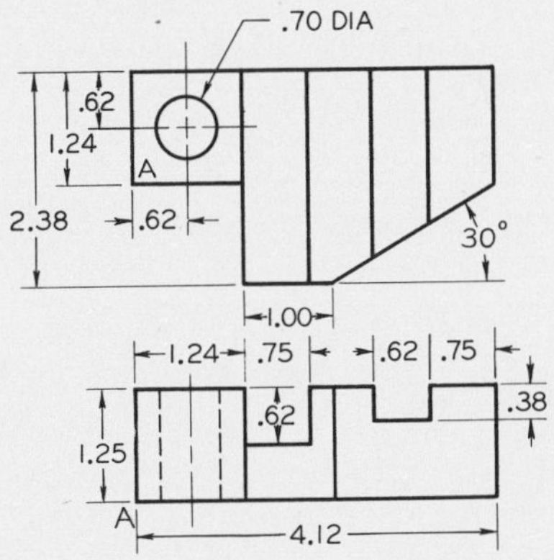
.70 DIA
.62
1.24
A
2.38
.62
30°
1.00
1.24
.75
.62
.75
.38
.62
1.25
A
4.12

(2)

A

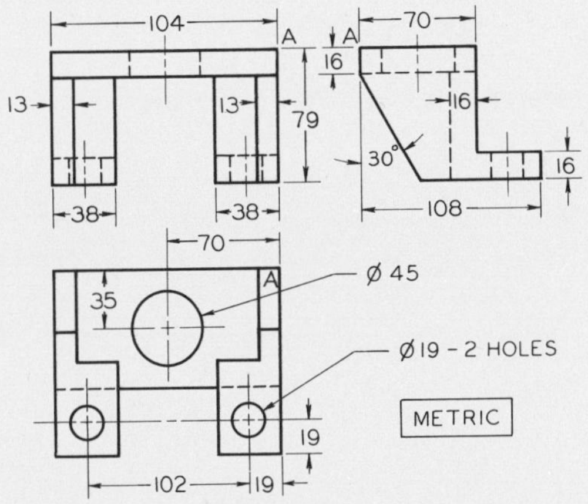
104
A
13
13
79
38
38
A
70
16
16
30°
16
108
70
A
35
Ø 45
Ø 19 - 2 HOLES
METRIC
19
102
19

BED SUPPORT

Draw isometric drawing.
Half size

SECTION	DRAWN BY:		SHEET 69

(1)

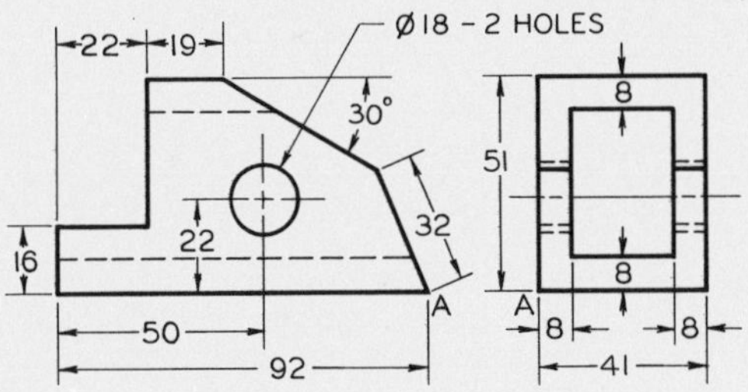
Ø18 - 2 HOLES
22
19
30°
51
8
32
22
16
A
50
92
41

METRIC

TIGHTENER

Draw isometric drawing.

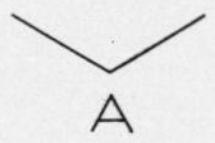
A

(2)

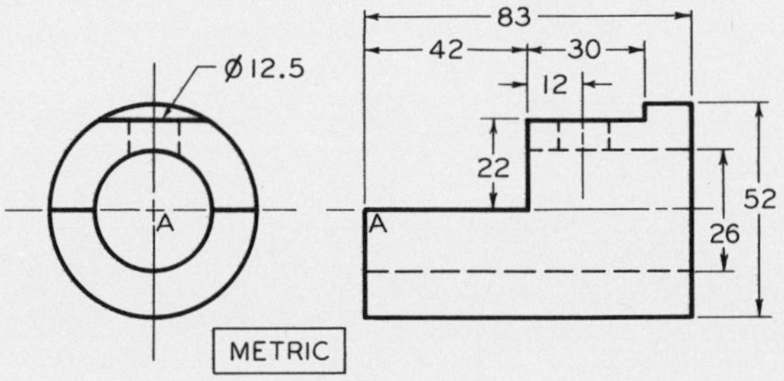
Ø12.5
A
83
42
30
12
22
52
26

METRIC

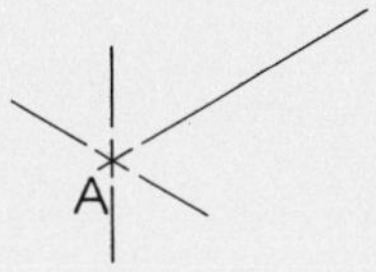
A

SLEEVE

Draw isometric drawing.

SECTION	DRAWN BY:		SHEET 70

(1)

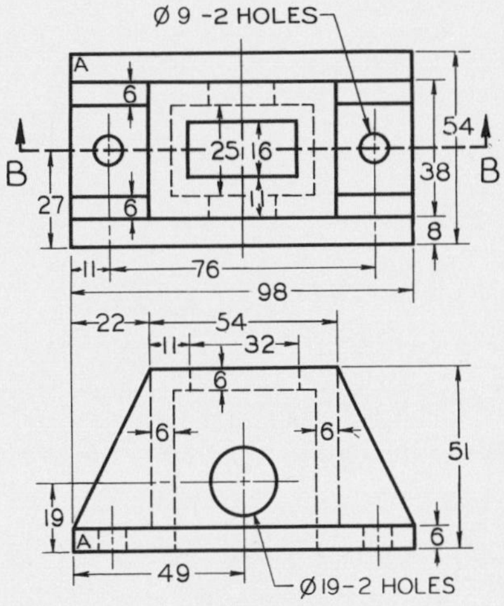
Ø 9 -2 HOLES
A
6
25
16
6
11
B
27
54
38
8
B
11
76
98
22
54
11
32
6
6
6
51
19
A
6
49
Ø 19-2 HOLES
METRIC

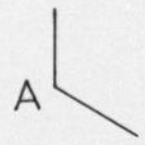
A

BASE

Draw isometric section B-B.

(2)

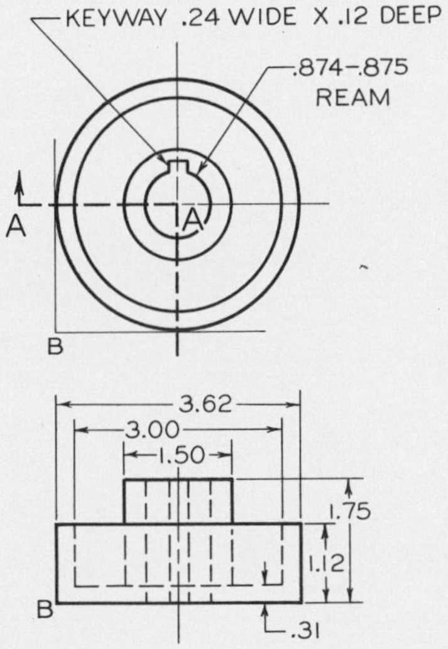
KEYWAY .24 WIDE X .12 DEEP
.874-.875
REAM
A
A
B
3.62
3.00
1.50
1.75
1.12
B
.31

FLANGE

Draw isometric half section A-A.

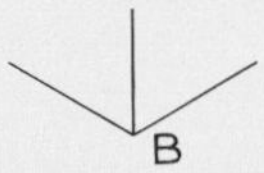
B

SECTION	DRAWN BY:		SHEET 72

(1)

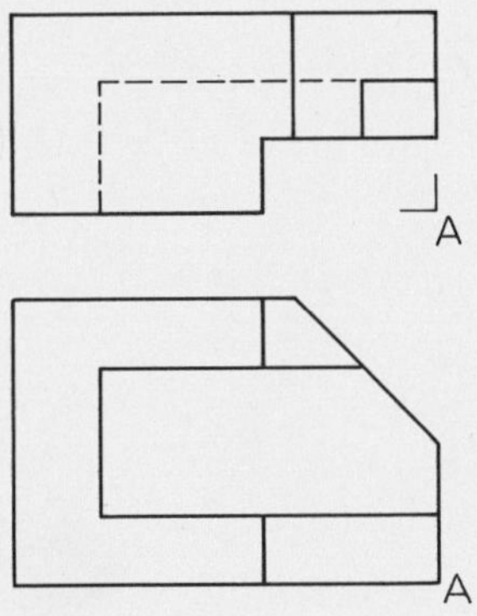
A
A

PLUNGER STOP

Complete the cavalier drawing.

A

(2)

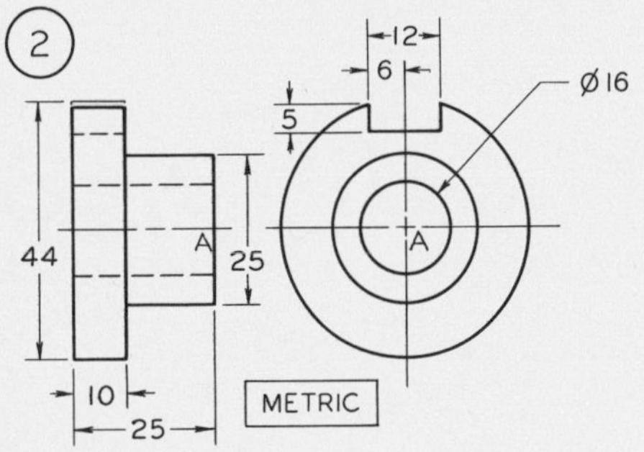
12
6
5
Ø16
44
A
25
A
10
25
METRIC

A

INDEX PLATE

Draw cabinet drawing.

(3)

BEARING

Draw cavalier drawing.

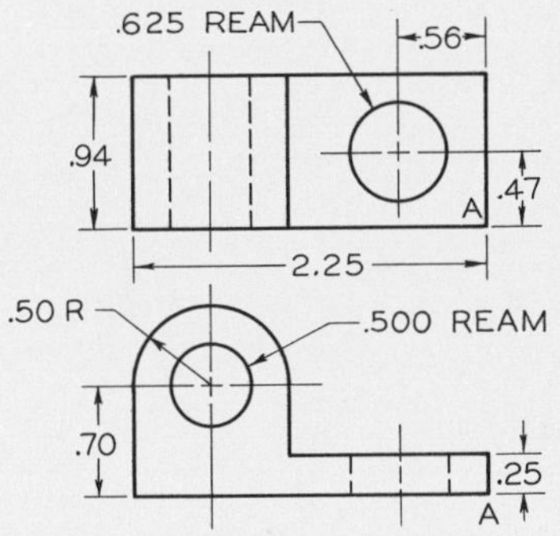
.625 REAM
.56
.94
A
.47
2.25
.50 R
.500 REAM
.70
.25
A

A

(4)

EJECTOR

Draw cavalier drawing.

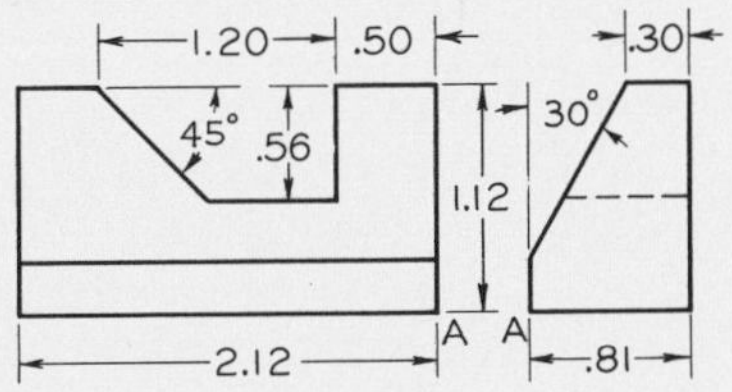
1.20
.50
.30
45°
.56
30°
1.12
A
A
2.12
.81

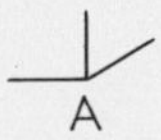
A

(5)

SUPPORT

Draw cabinet drawing.

METRIC

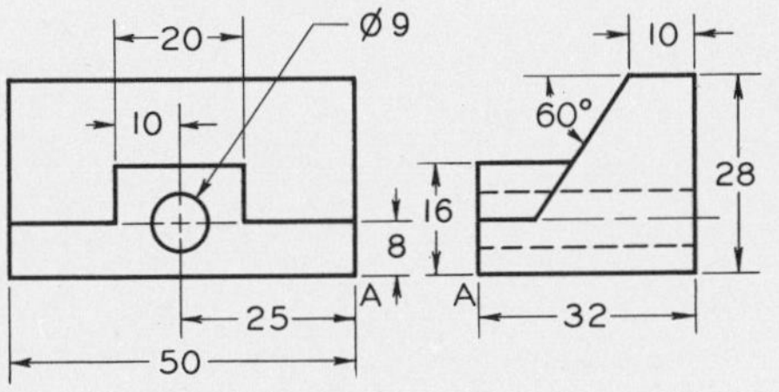
20
Ø9
10
10
60°
28
16
8
A
A
25
32
50

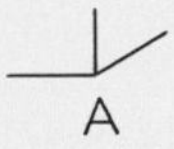
A

(6)

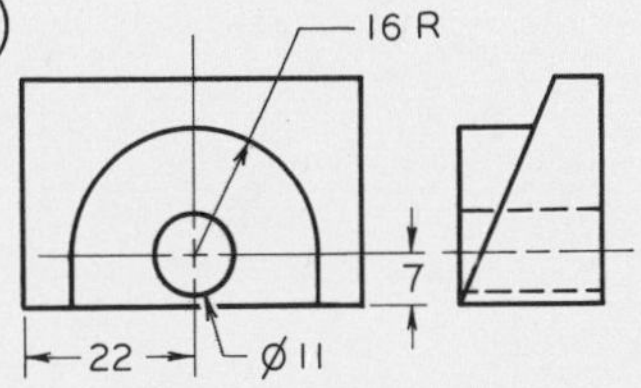
16 R
7
22
Ø11

CLAMP JAW

Complete the cavalier drawing.

METRIC

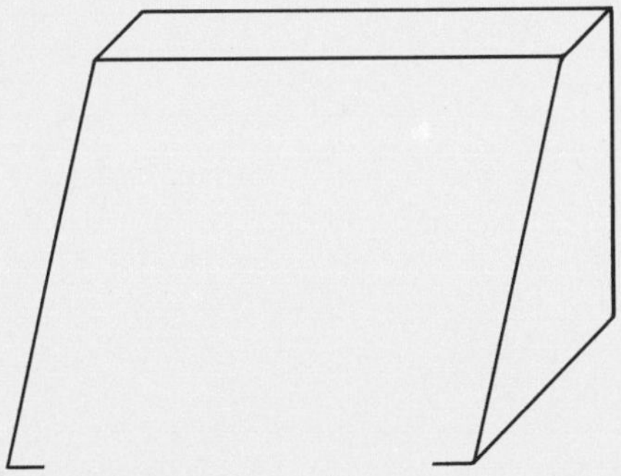

SECTION	DRAWN BY:		SHEET 74

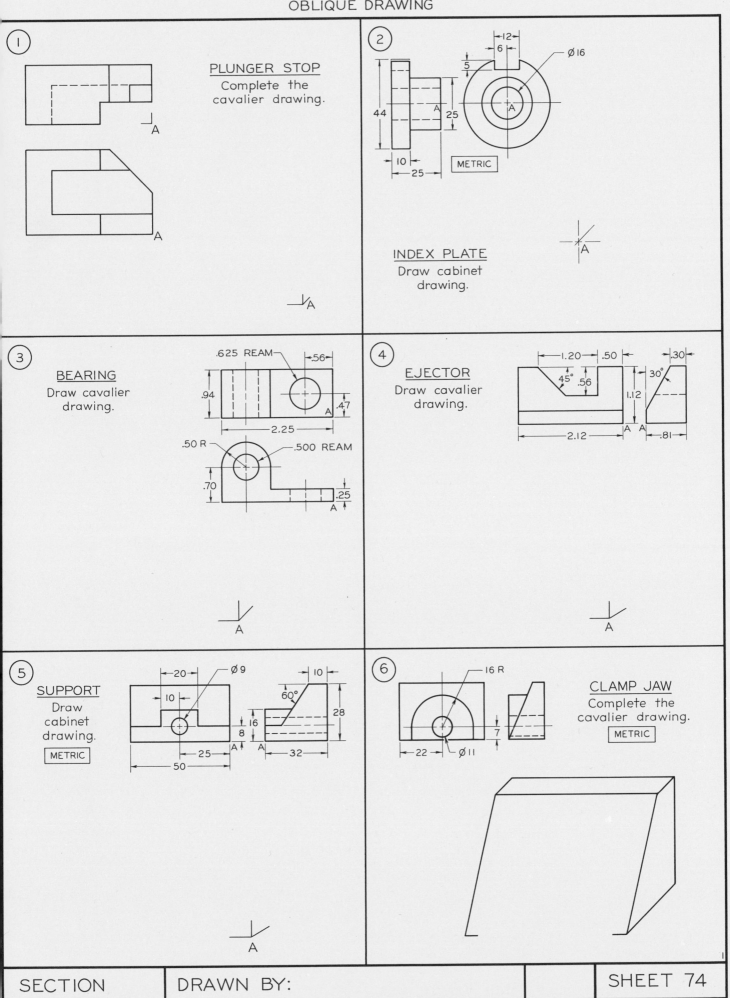

(1)

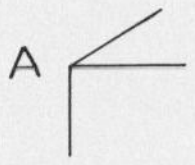
A

B

STAND BASE

Draw cavalier drawing.

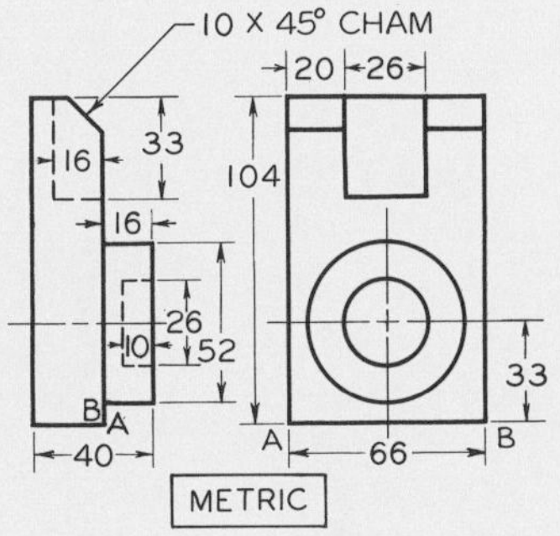
10 X 45° CHAM
20
26
16
33
104
16
26
10
52
33
B
A
A
B
40
66
METRIC

(2)

STOP LINK

Draw cabinet drawing.

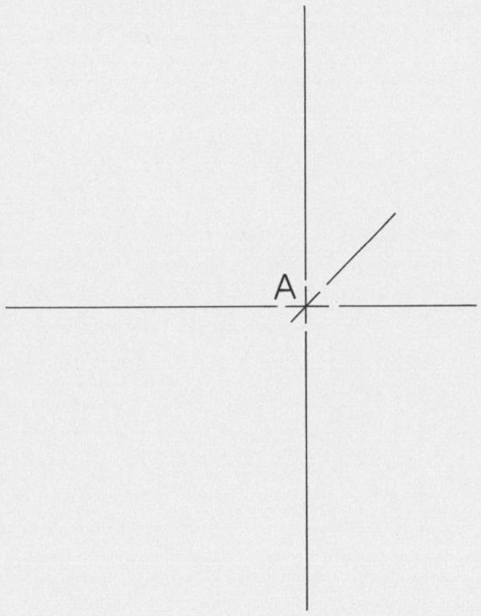
A

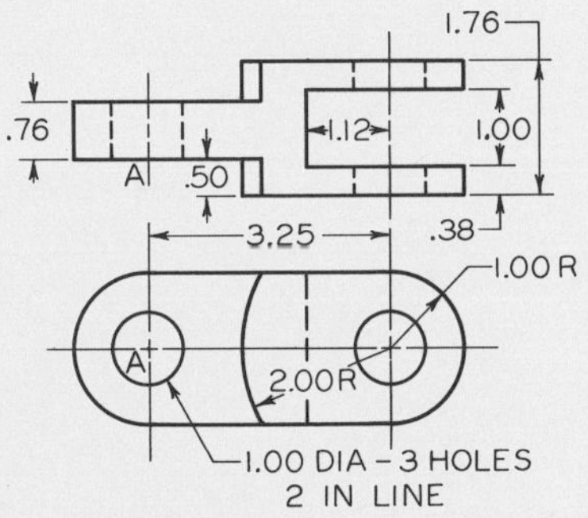
1.76
.76
1.12
1.00
A
.50
.38
3.25
1.00 R
A
2.00R
1.00 DIA - 3 HOLES
2 IN LINE

SECTION	DRAWN BY:		SHEET 75

(1)

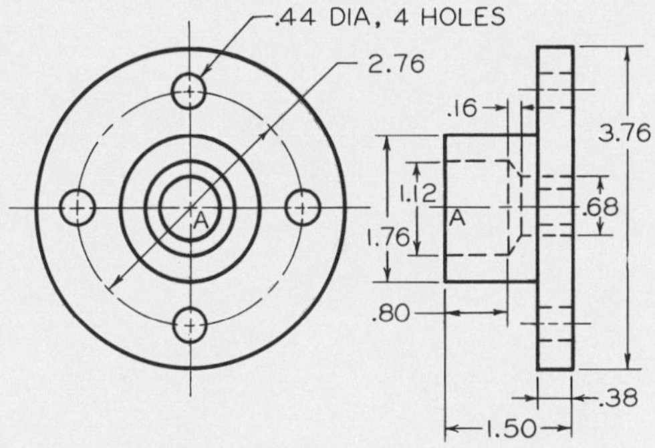
.44 DIA, 4 HOLES
2.76
A
.16
3.76
1.12
A
.68
1.76
.80
.38
1.50

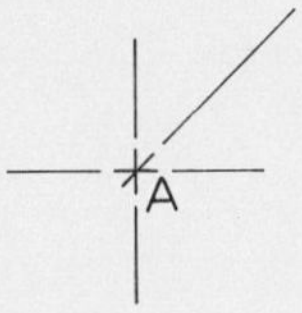
A

FLANGE

Draw cabinet drawing in half section.

(2)

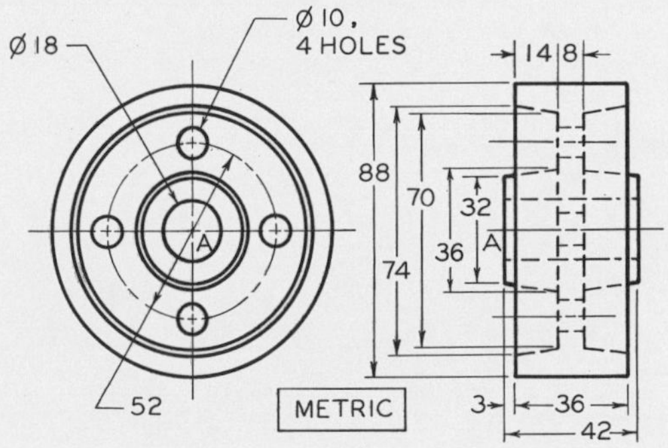
Ø18
Ø10,
4 HOLES
A
52
METRIC
14
8
88
70
32
36
74
A
3
36
42

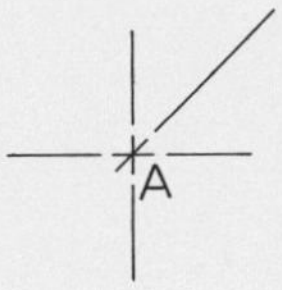
A

TRUCK WHEEL

Draw cavalier drawing in full section.

SECTION	DRAWN BY:		SHEET 77

(1)

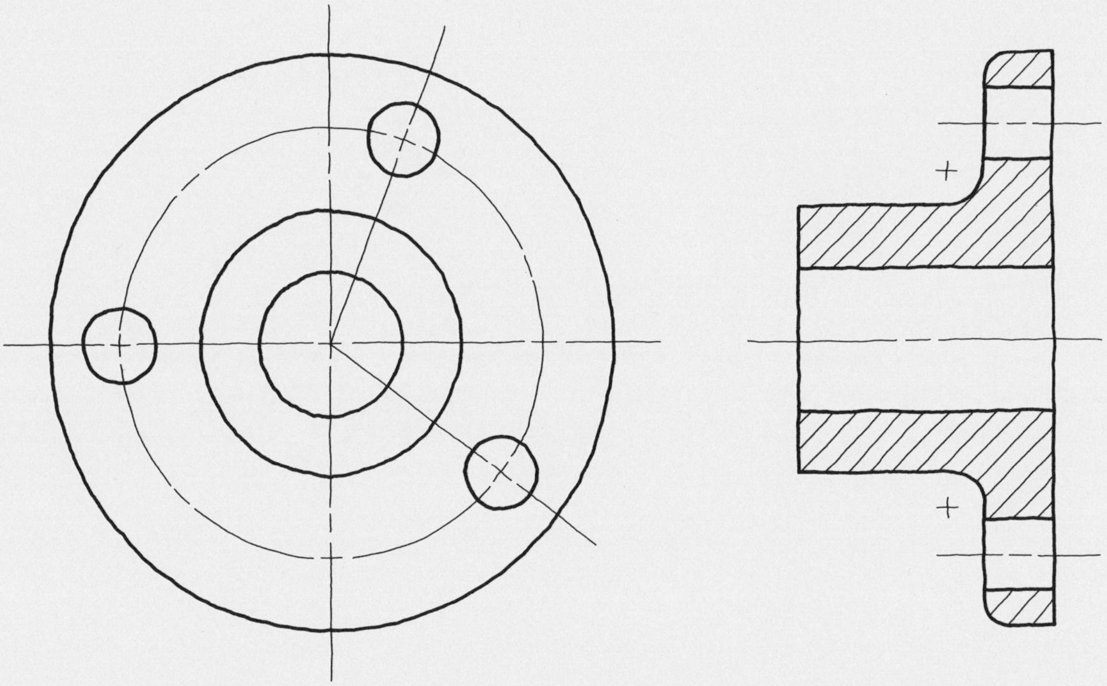

SCALE : 1 = 1

(2)

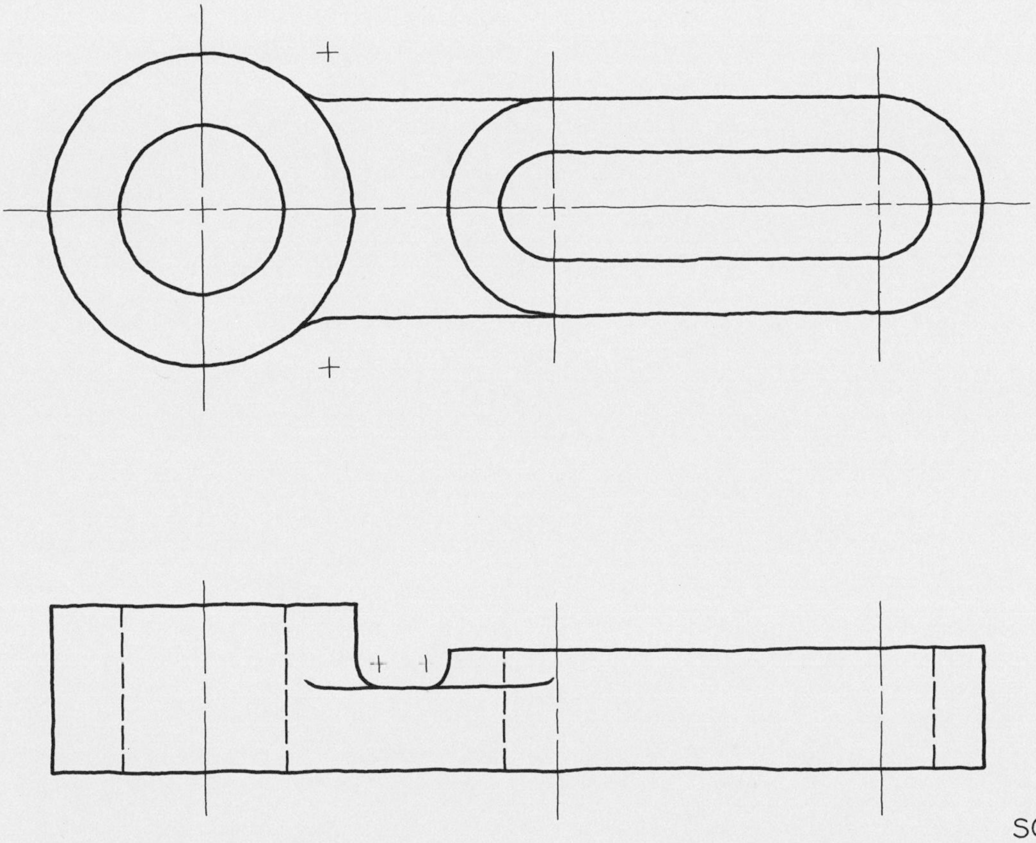

SCALE : 1 = 2

SECTION	DRAWN BY:		SHEET 78

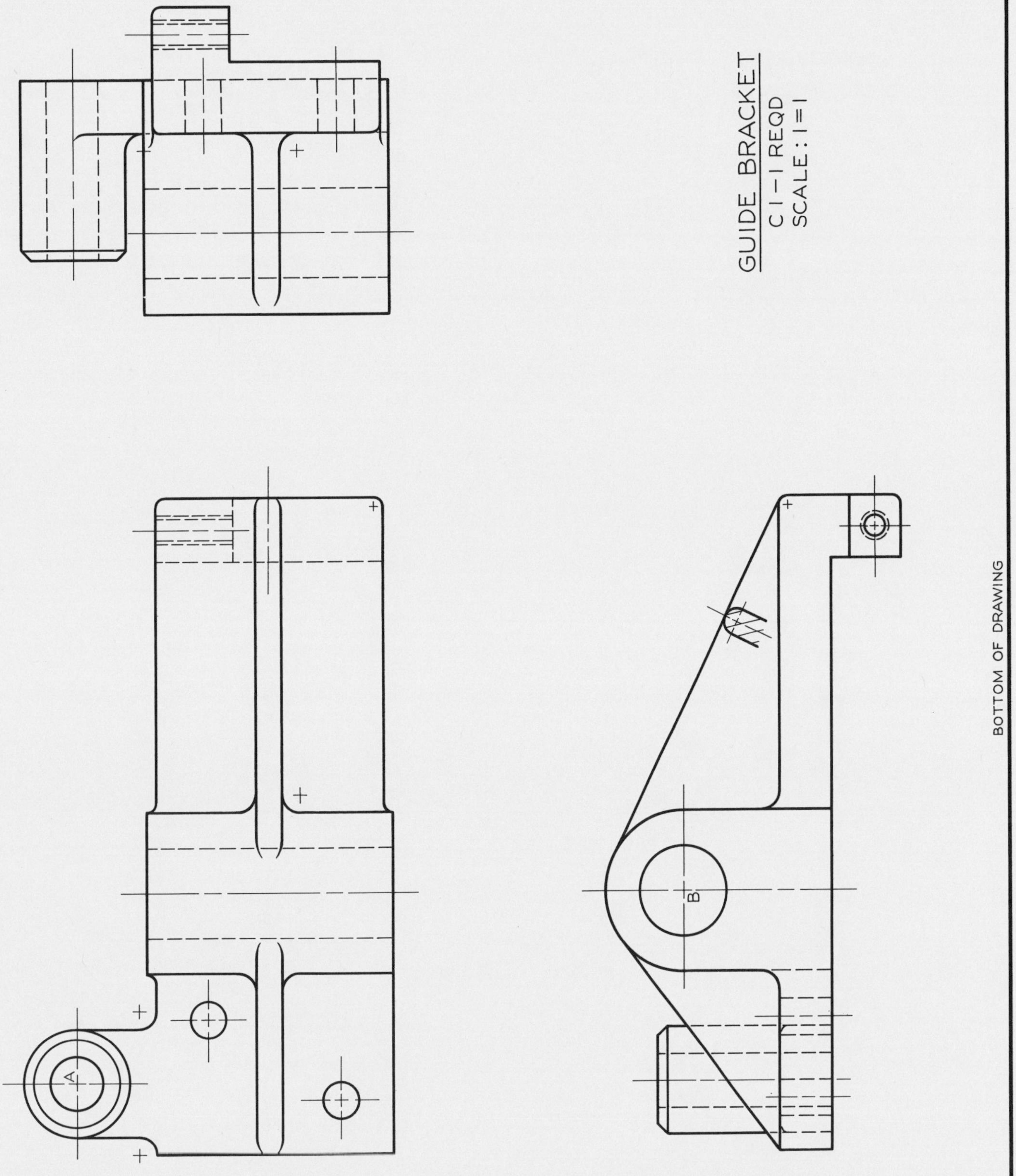
GUIDE BRACKET
C1-1 REQD
SCALE: 1=1
A
B
BOTTOM OF DRAWING

SECTION	DRAWN BY:		SHEET 81

(1) Show 3 ways of giving tolerances, using the basic hole system.

Limits for hole are: 4.750 and 4.754
Limits fcr shaft are: 4.742 and 4.746

1st method (limits):

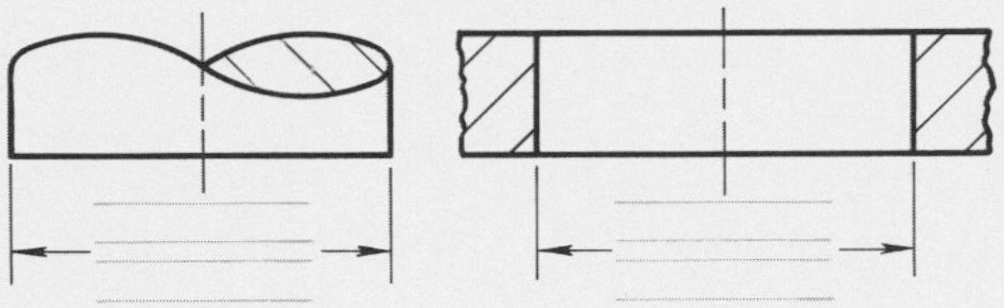

2nd method (plus and minus):

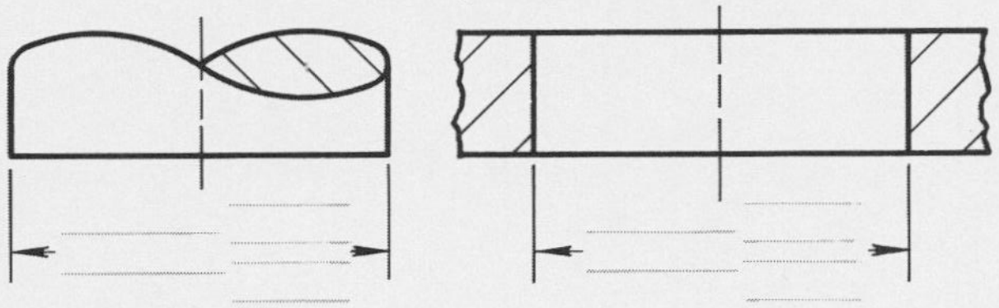

2nd method (plus or minus):

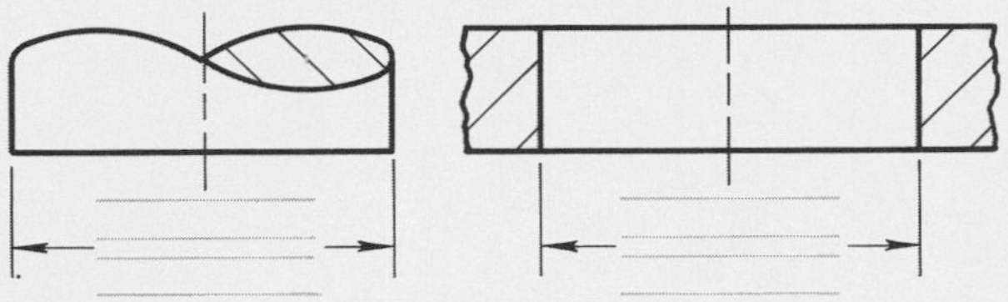

(2) Dimension the distance between the holes, using bilateral tolerance (equal ±).

B = 1.625 T = .006

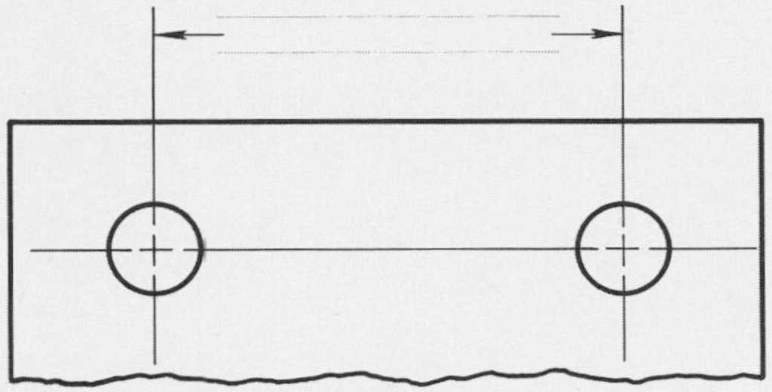

(3) Give limits with unilateral tolerances, using the basic hole system.
B = 1.875, A = .003 (positive), TH = TS = .002

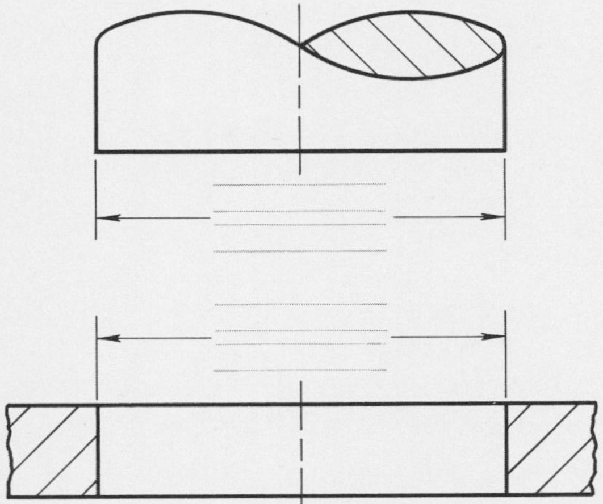

Check the correct fit:
This is a clearance ( ), transition ( ), interference ( ) fit.

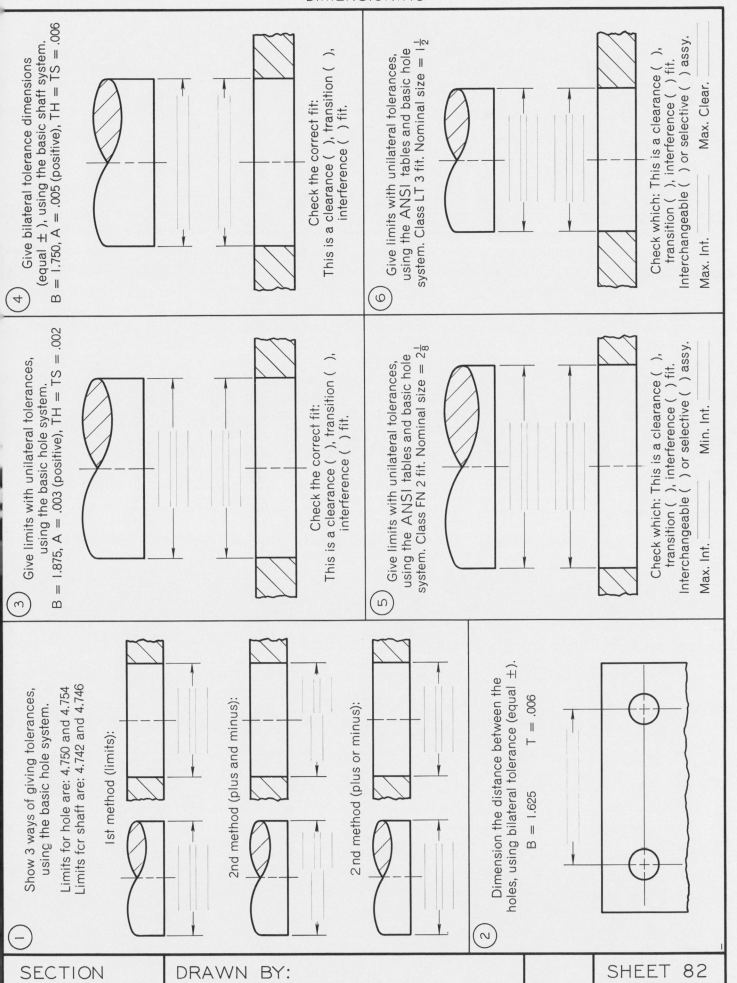

(4) Give bilateral tolerance dimensions (equal ±), using the basic shaft system.
B = 1.750, A = .005 (positive), TH = TS = .006

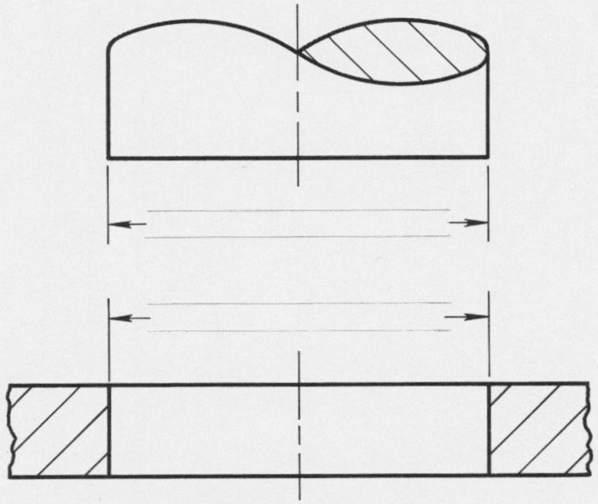

Check the correct fit:
This is a clearance ( ), transition ( ), interference ( ) fit.

(5) Give limits with unilateral tolerances, using the ANSI tables and basic hole system. Class FN 2 fit. Nominal size = $2\frac{1}{8}$

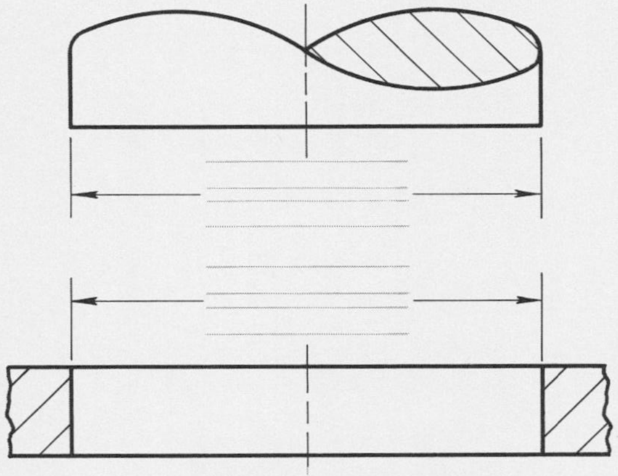

Check which: This is a clearance ( ), transition ( ), interference ( ) fit.
Interchangeable ( ) or selective ( ) assy.

Max. Int. ______ Min. Int. ______

(6) Give limits with unilateral tolerances, using the ANSI tables and basic hole system. Class LT 3 fit. Nominal size = $1\frac{1}{2}$

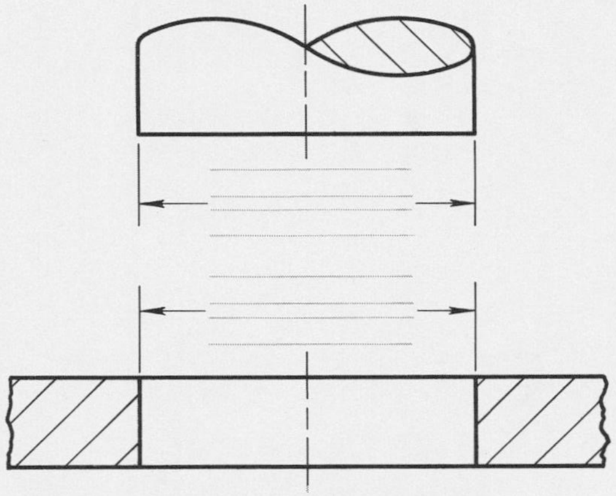

Check which: This is a clearance ( ), transition ( ), interference ( ) fit.
Interchangeable ( ) or selective ( ) assy.

Max. Int. ______ Max. Clear. ______

SECTION	DRAWN BY:		SHEET 82

① BRACKET
FOR
TRUCK WHEEL
CAST IRON – 1 REQD
SCALE: 1=1

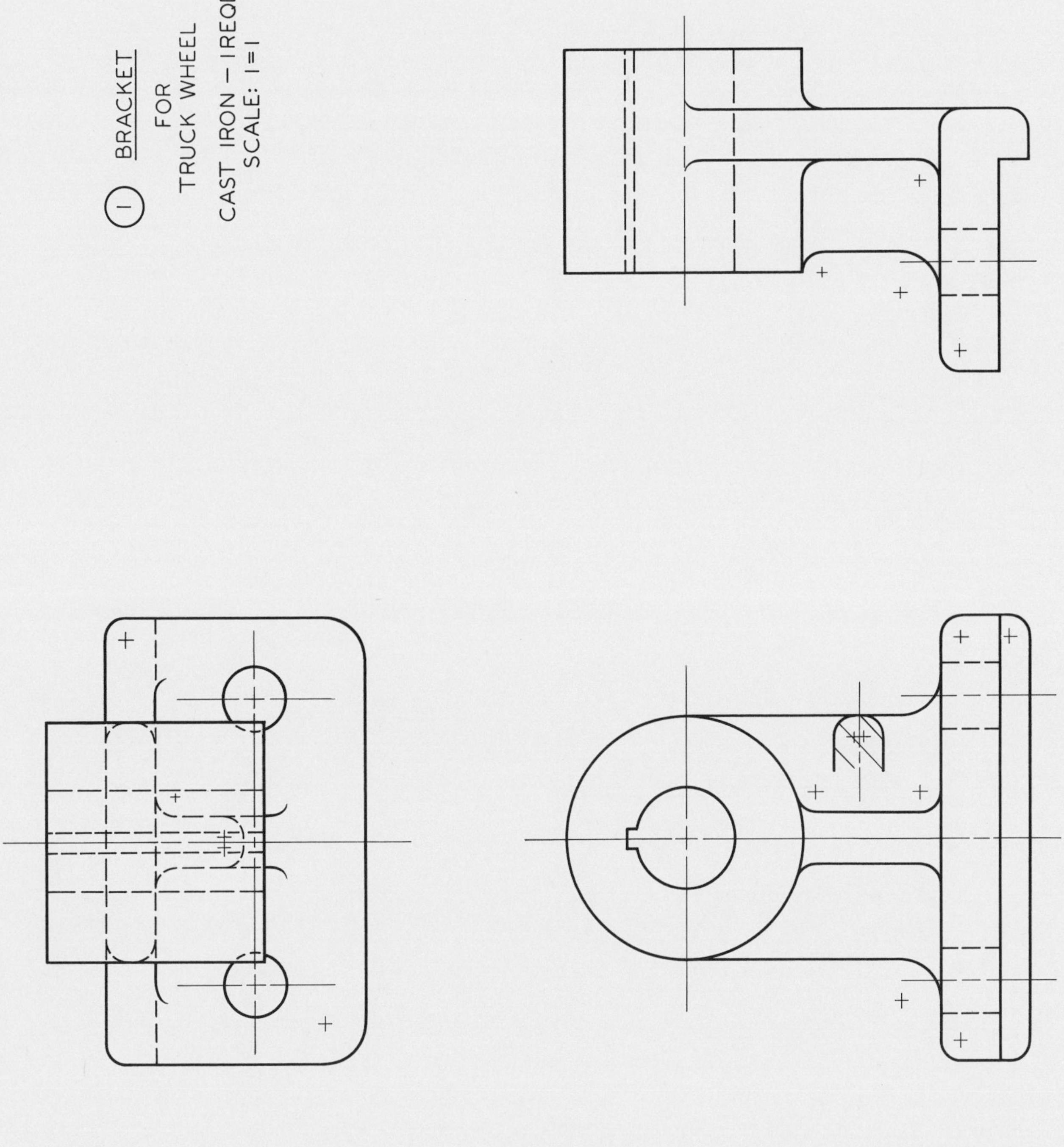

BOTTOM OF DRAWING

SECTION	DRAWN BY:		SHEET 83

STANDARD PARTS

(4) I - HEX NUT, FIN. - ____

(5) I - LOCK WASHER, REG - ____

(6) I - WOODRUFF KEY, NO. 405

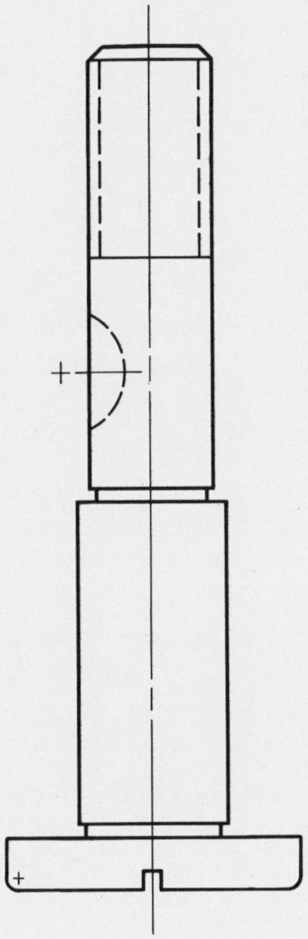

(3) WHEEL SCREW
CRS - I REQD

SCALE: I=I

BOTTOM OF DRAWING

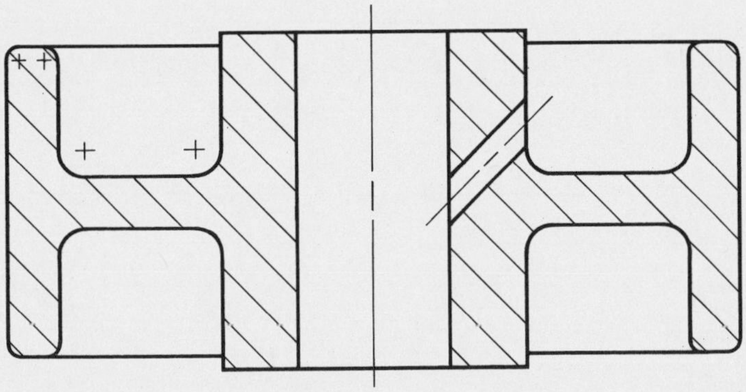

(2) WHEEL
CAST IRON - I REQD

SECTION	DRAWN BY:		SHEET 84

THREADS & FASTENERS

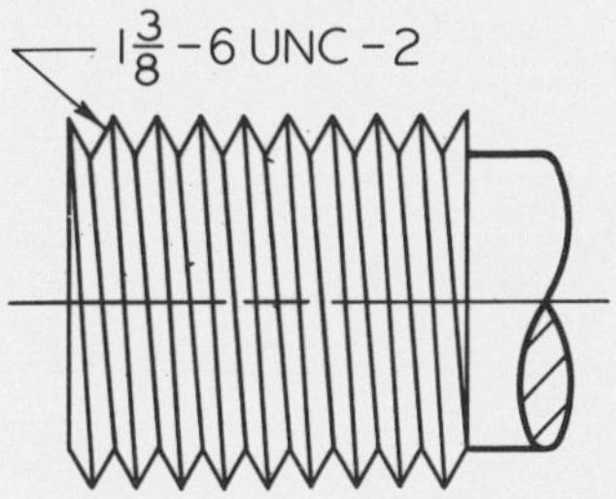

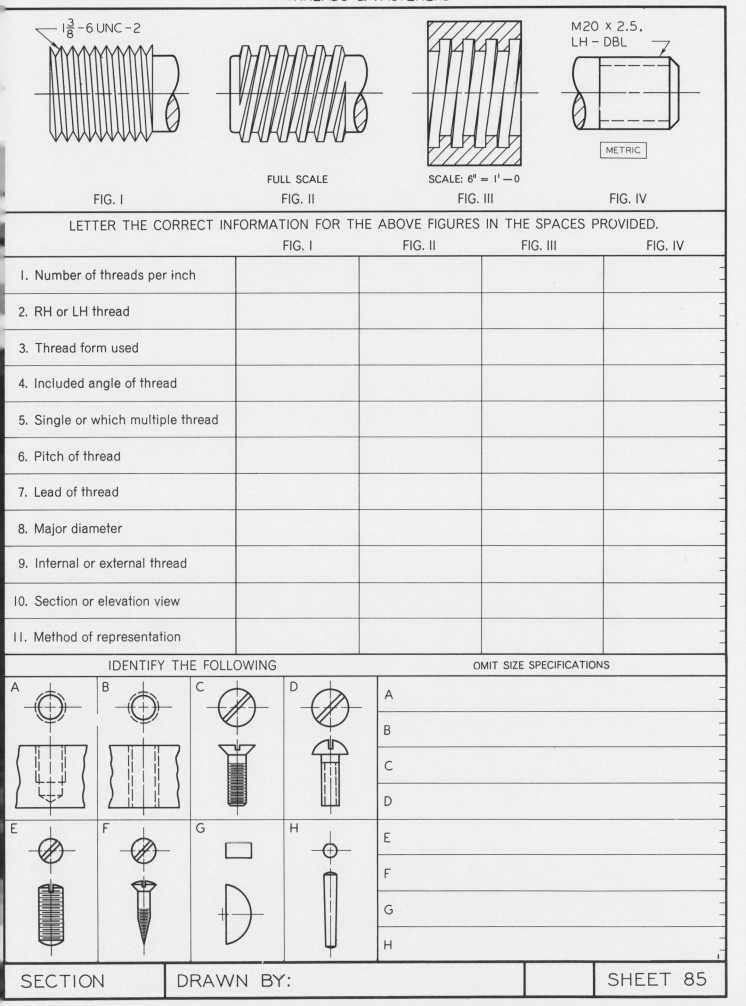

$1\frac{3}{8}$ - 6 UNC - 2

FIG. I

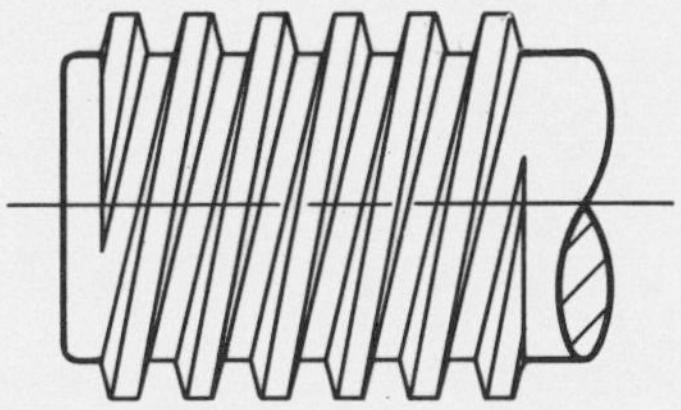

FULL SCALE

FIG. II

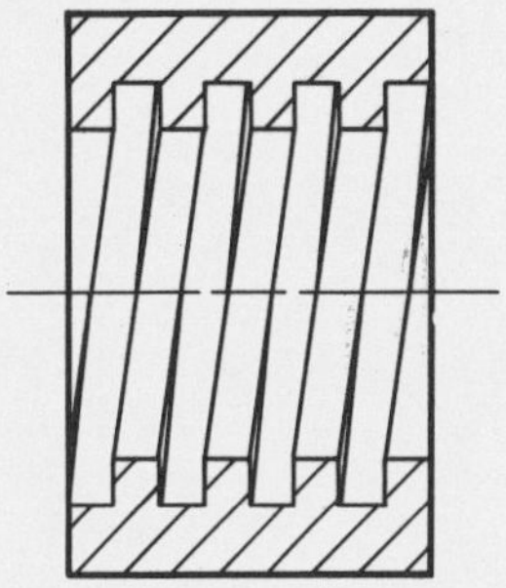

SCALE: 6" = 1' — 0

FIG. III

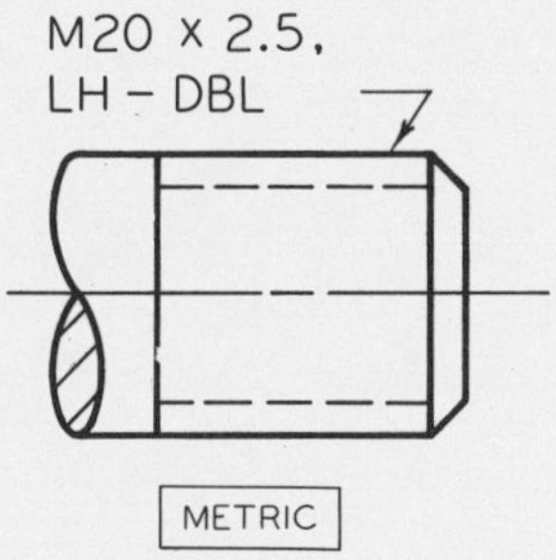

M20 x 2.5,
LH - DBL
METRIC

FIG. IV

LETTER THE CORRECT INFORMATION FOR THE ABOVE FIGURES IN THE SPACES PROVIDED.

	FIG. I	FIG. II	FIG. III	FIG. IV
1. Number of threads per inch				
2. RH or LH thread				
3. Thread form used				
4. Included angle of thread				
5. Single or which multiple thread				
6. Pitch of thread				
7. Lead of thread				
8. Major diameter				
9. Internal or external thread				
10. Section or elevation view				
11. Method of representation				

IDENTIFY THE FOLLOWING

OMIT SIZE SPECIFICATIONS

A
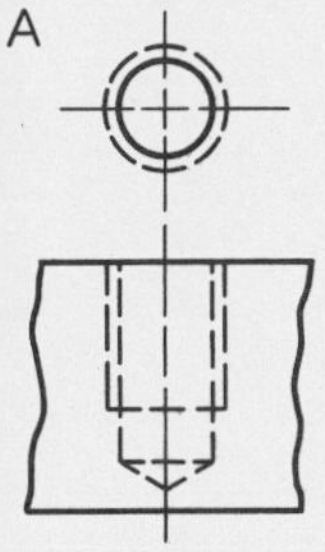

B
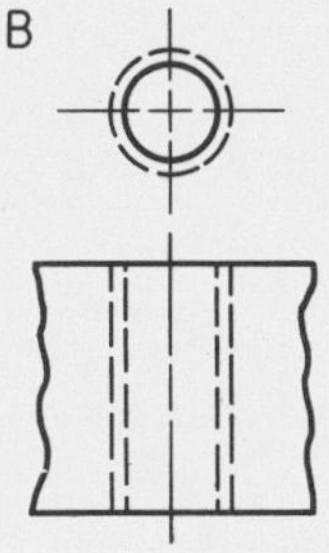

C
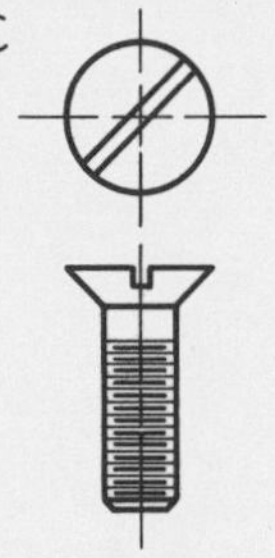

D
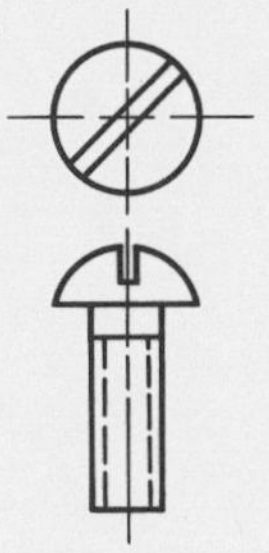

E
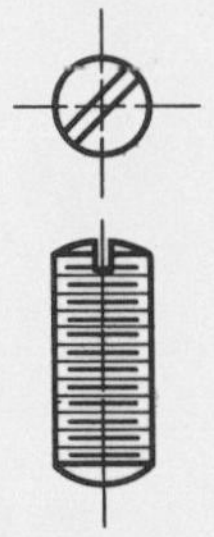

F
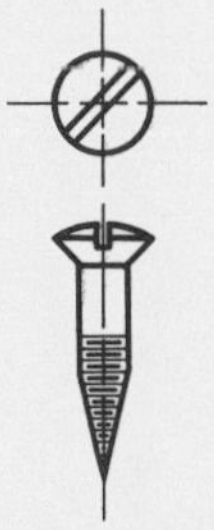

G

H

A	
B	
C	
D	
E	
F	
G	
H	

SECTION	DRAWN BY:		SHEET 85

(1) Draw Metric Coarse threads. Give thread note.

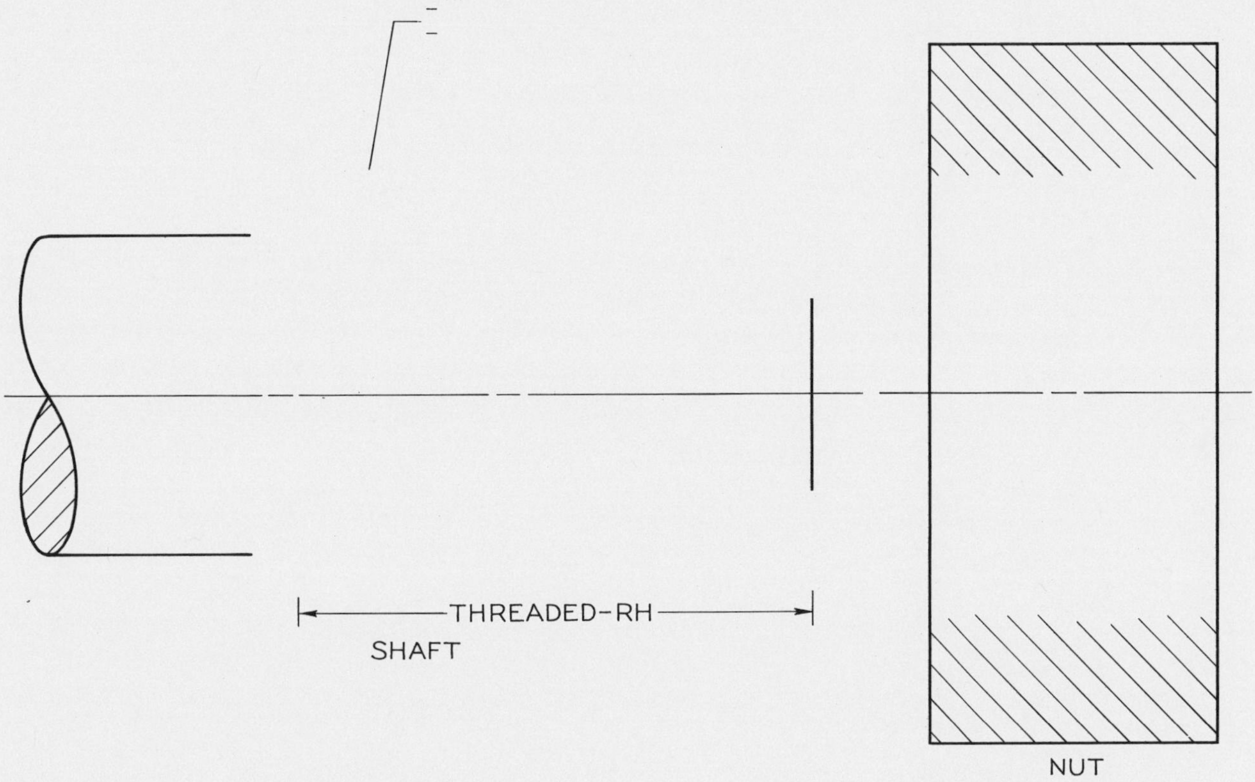
THREADED-RH
SHAFT
NUT

(2) Draw Unified Coarse threads. Give thread note.

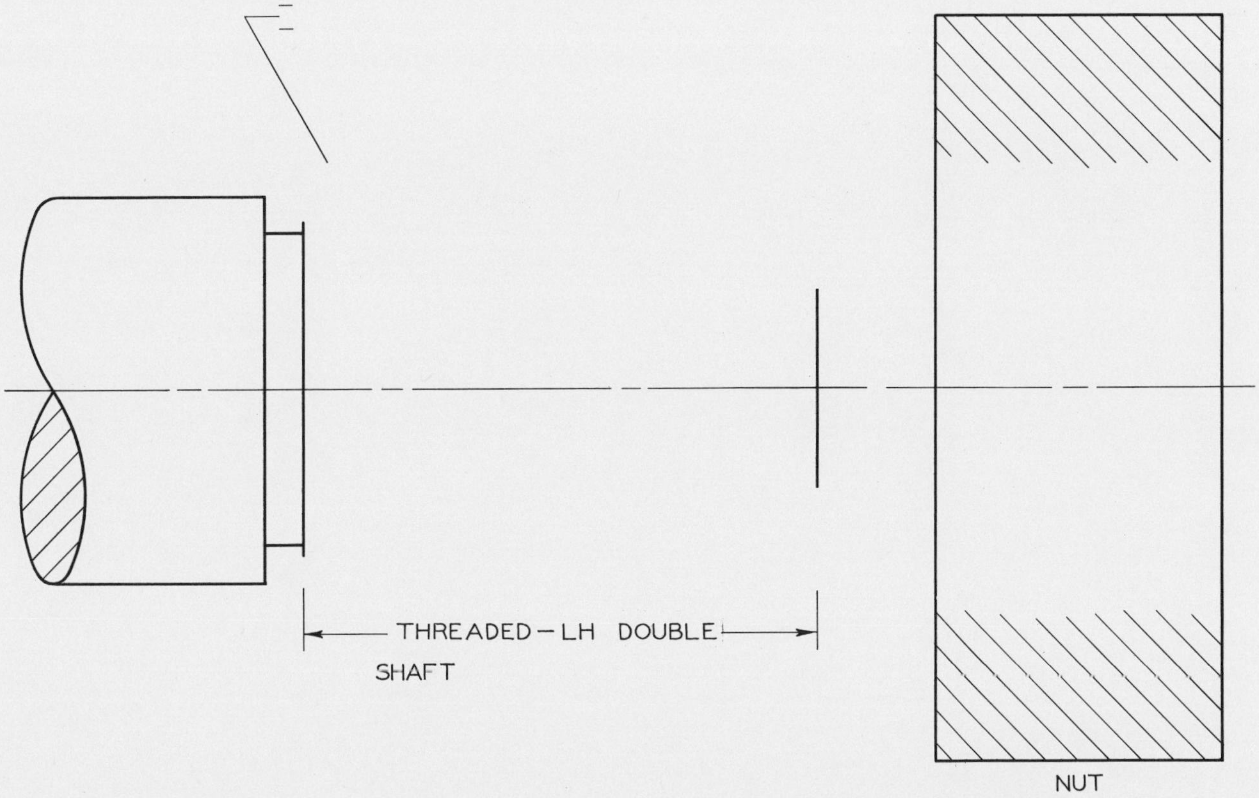
THREADED – LH DOUBLE
SHAFT
NUT

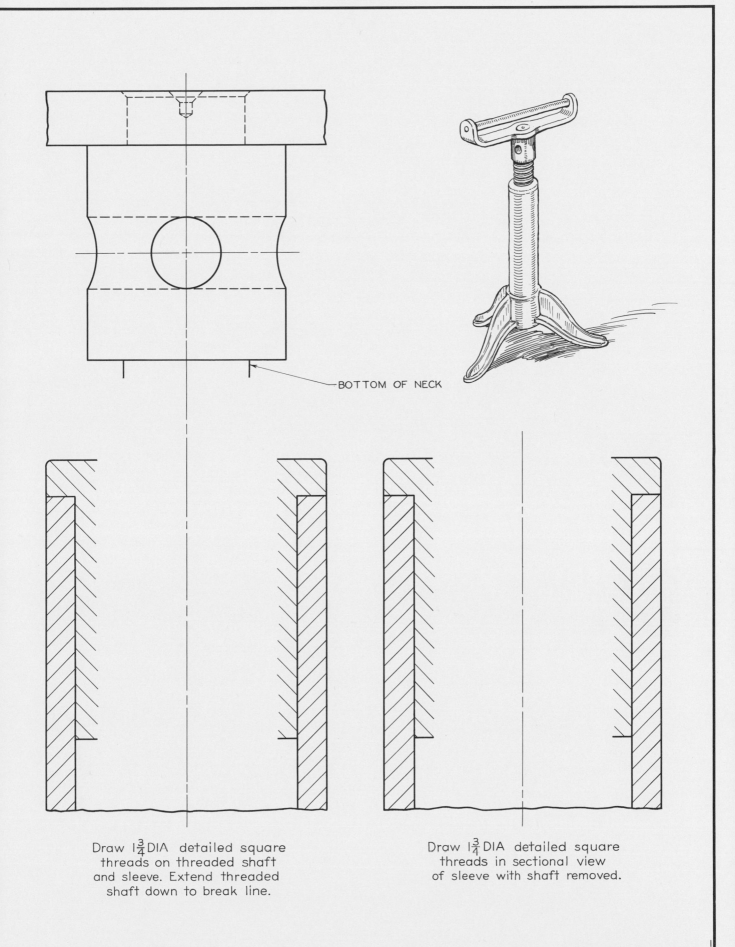

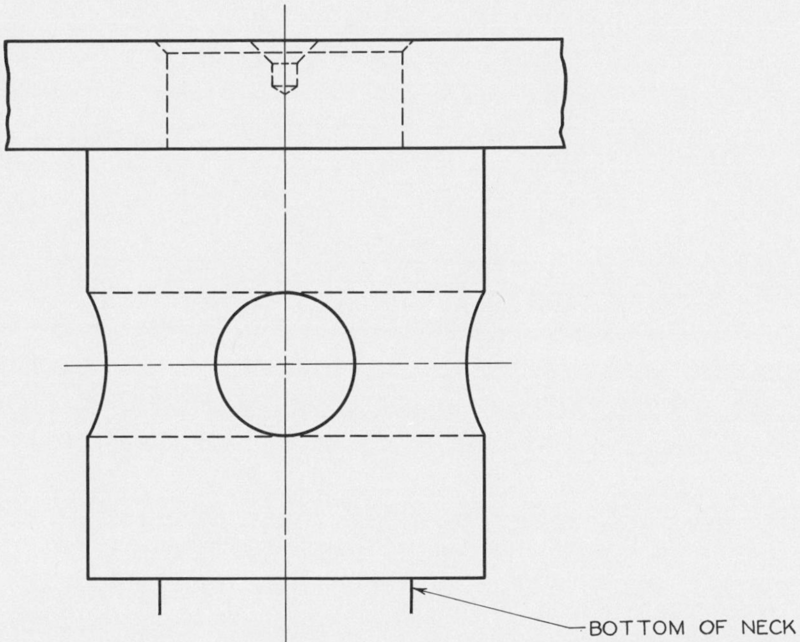
BOTTOM OF NECK

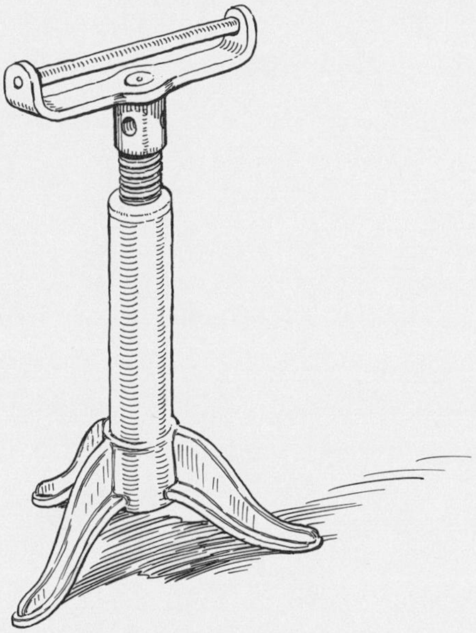

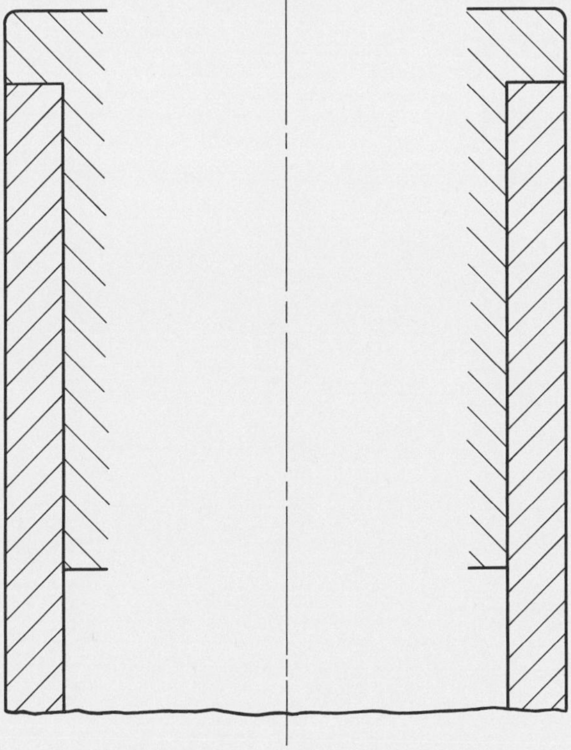

Draw $1\frac{3}{4}$ DIA detailed square threads on threaded shaft and sleeve. Extend threaded shaft down to break line.

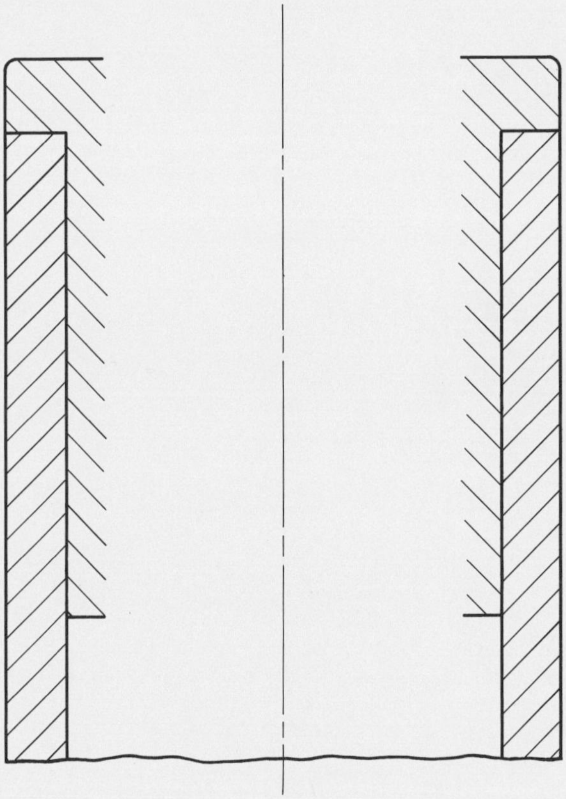

Draw $1\frac{3}{4}$ DIA detailed square threads in sectional view of sleeve with shaft removed.

SECTION	DRAWN BY:		SHEET 87

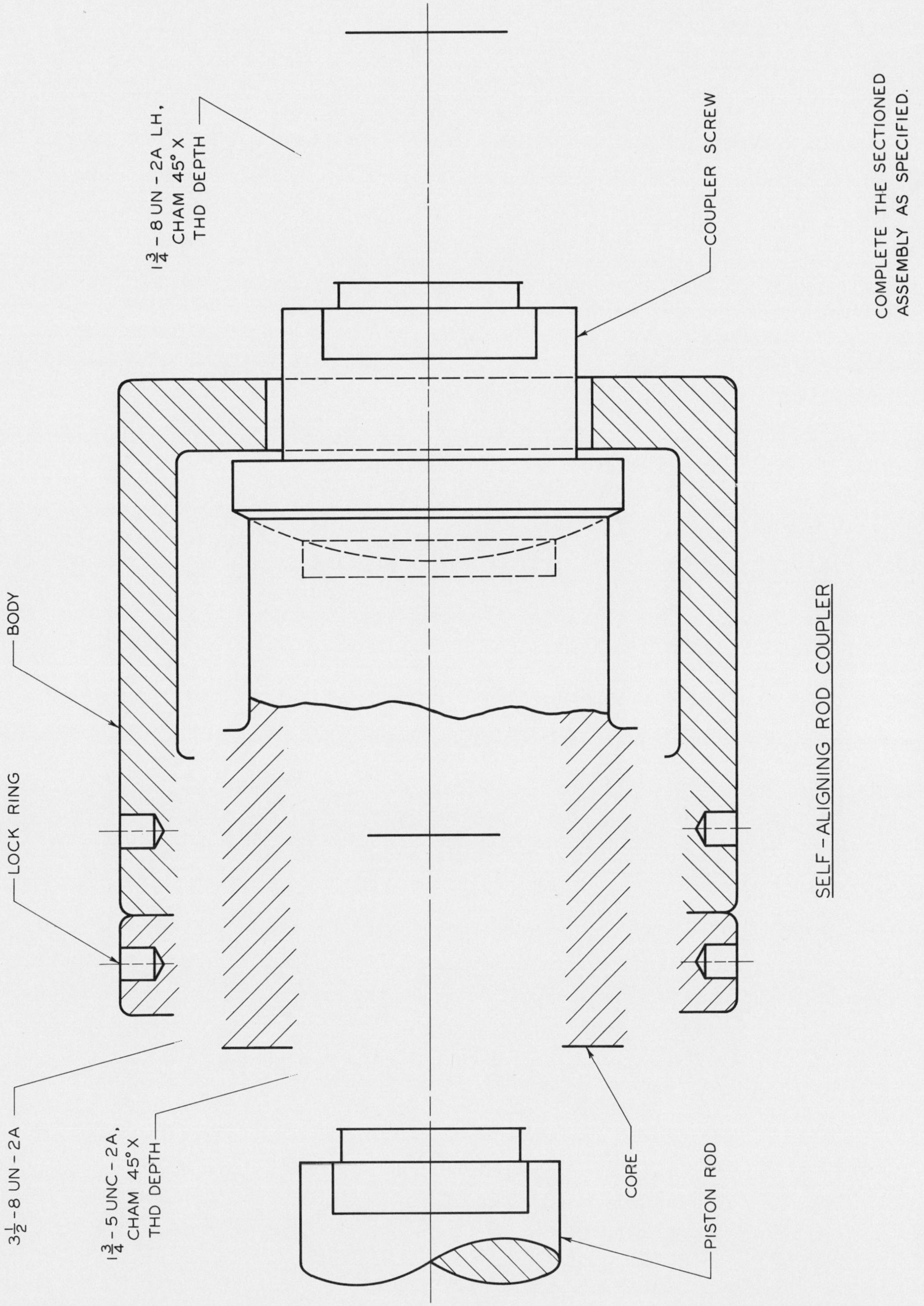
COMPLETE THE SECTIONED ASSEMBLY AS SPECIFIED.
SELF-ALIGNING ROD COUPLER
$1\frac{3}{4}$ - 8 UN - 2A LH, CHAM 45° X THD DEPTH
COUPLER SCREW
BODY
LOCK RING
$3\frac{1}{2}$ - 8 UN - 2A
$1\frac{3}{4}$ - 5 UNC - 2A, CHAM 45° X THD DEPTH
CORE
PISTON ROD

SECTION	DRAWN BY:		SHEET 89

COMPLETE THE VIEWS.

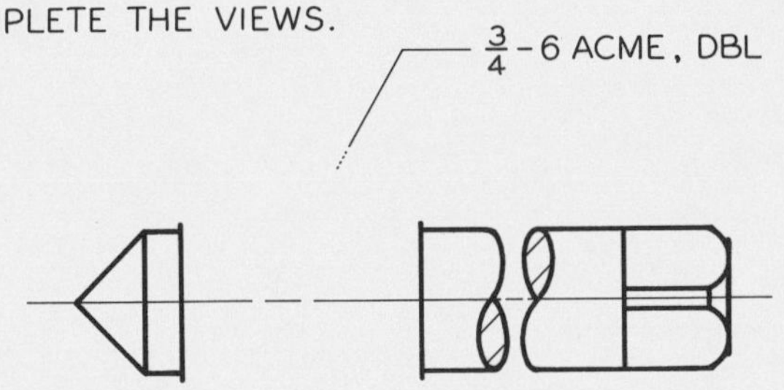
3/4-6 ACME, DBL

1. VALVE STEM

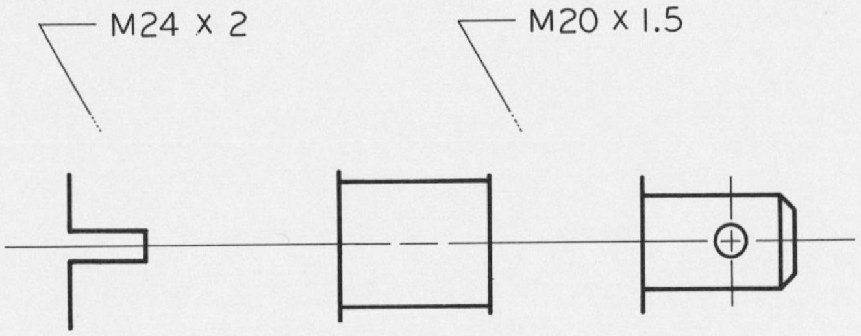
M24 X 2
M20 X 1.5

2. METERING SCREW

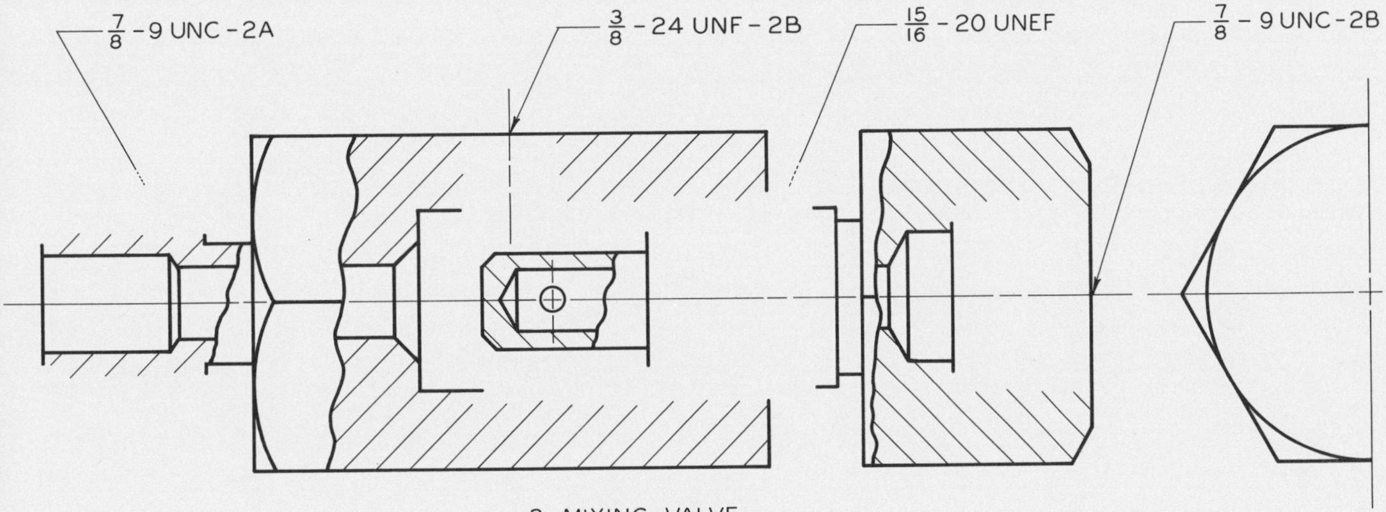
7/8-9 UNC-2A
3/8-24 UNF-2B
15/16-20 UNEF
7/8-9 UNC-2B

3. MIXING VALVE

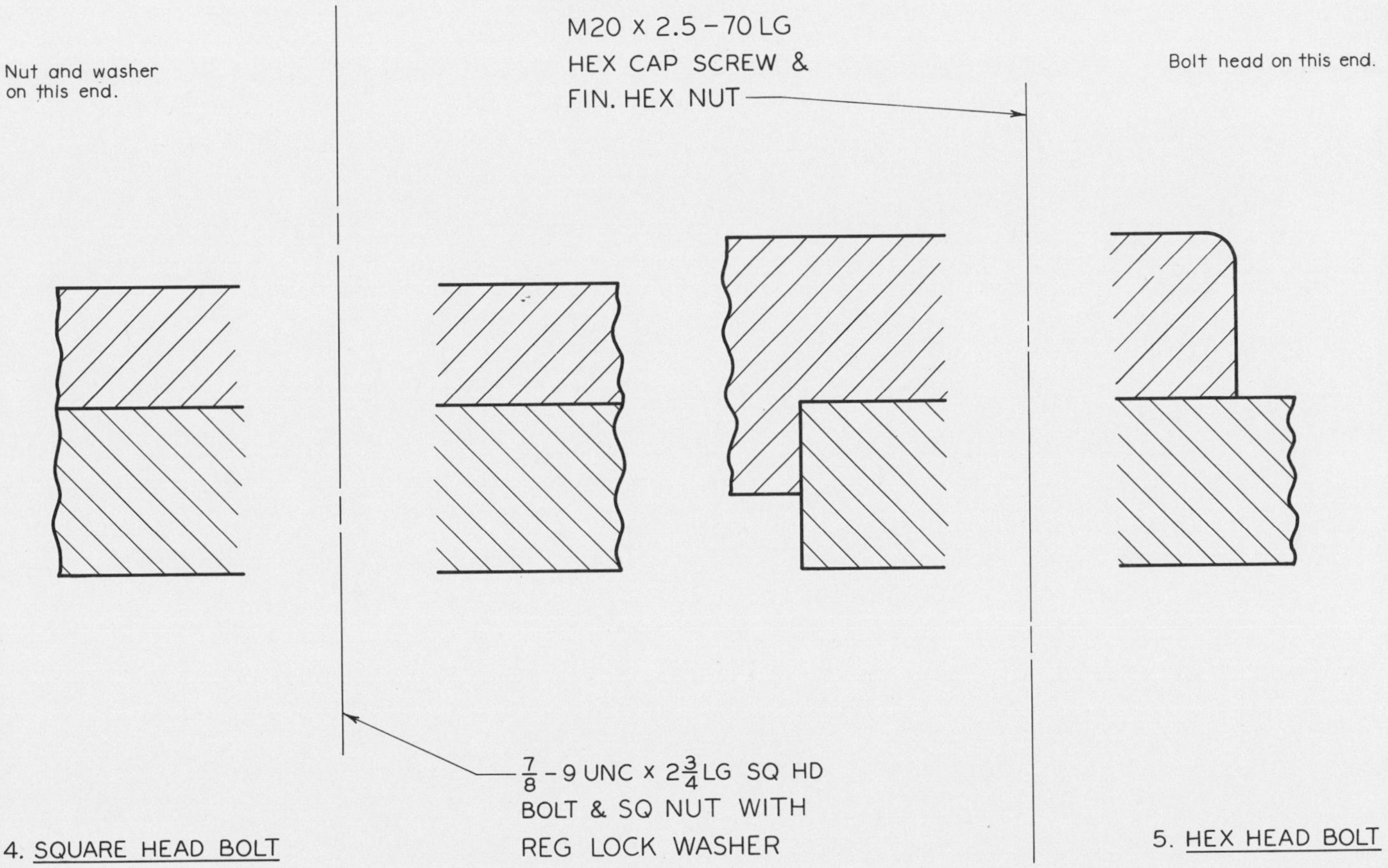
Nut and washer on this end.
M20 X 2.5-70 LG
HEX CAP SCREW &
FIN. HEX NUT
Bolt head on this end.
7/8-9 UNC X 2 3/4 LG SQ HD
BOLT & SQ NUT WITH
REG LOCK WASHER

4. SQUARE HEAD BOLT

5. HEX HEAD BOLT

SECTION	DRAWN BY:		SHEET 90

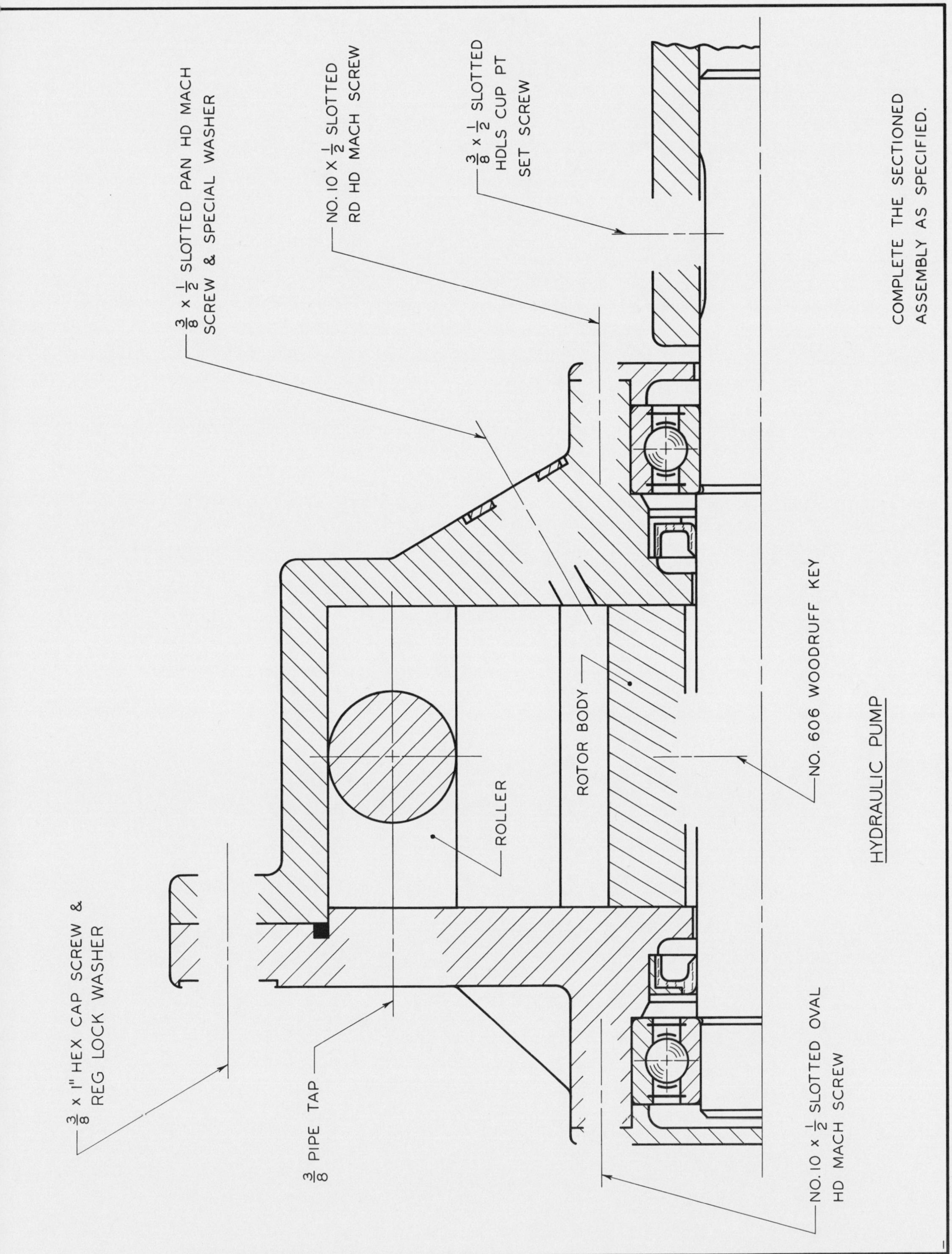
3/8 x 1/2 SLOTTED PAN HD MACH SCREW & SPECIAL WASHER
NO. 10 x 1/2 SLOTTED RD HD MACH SCREW
3/8 x 1/2 SLOTTED HDLS CUP PT SET SCREW
COMPLETE THE SECTIONED ASSEMBLY AS SPECIFIED.
NO. 606 WOODRUFF KEY
ROTOR BODY
ROLLER
HYDRAULIC PUMP
3/8 x 1" HEX CAP SCREW & REG LOCK WASHER
3/8 PIPE TAP
NO. 10 x 1/2 SLOTTED OVAL HD MACH SCREW

SECTION	DRAWN BY:		SHEET 91

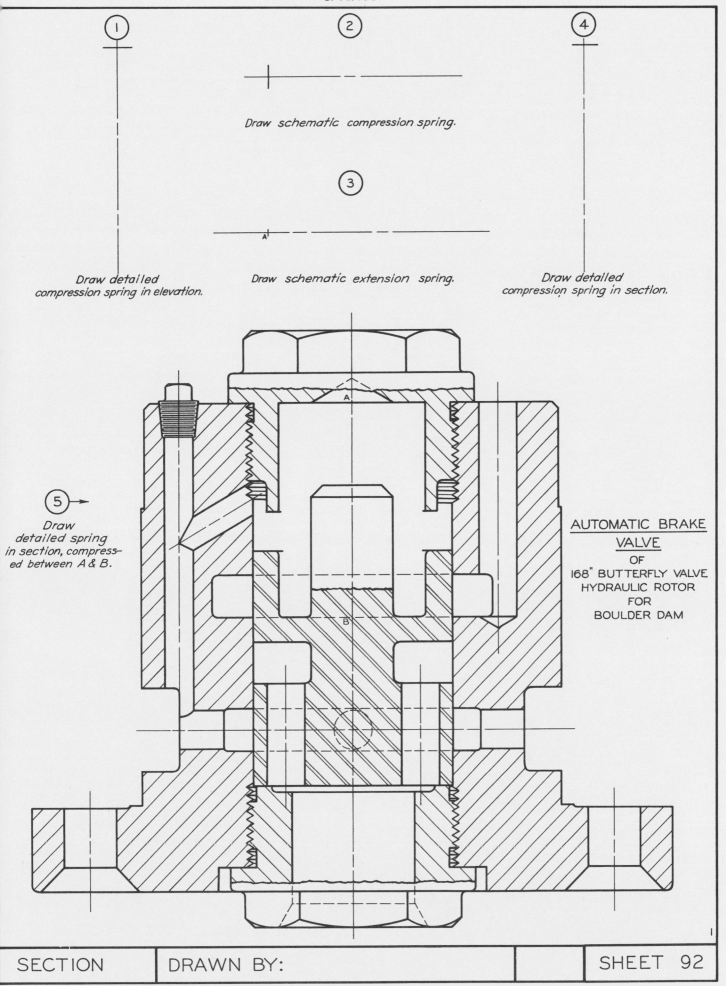

(1) Draw detailed compression spring in elevation.

(2) Draw schematic compression spring.

(3) Draw schematic extension spring.

A

(4) Draw detalled compression spring in section.

(5) → Draw detailed spring in section, compressed between A & B.

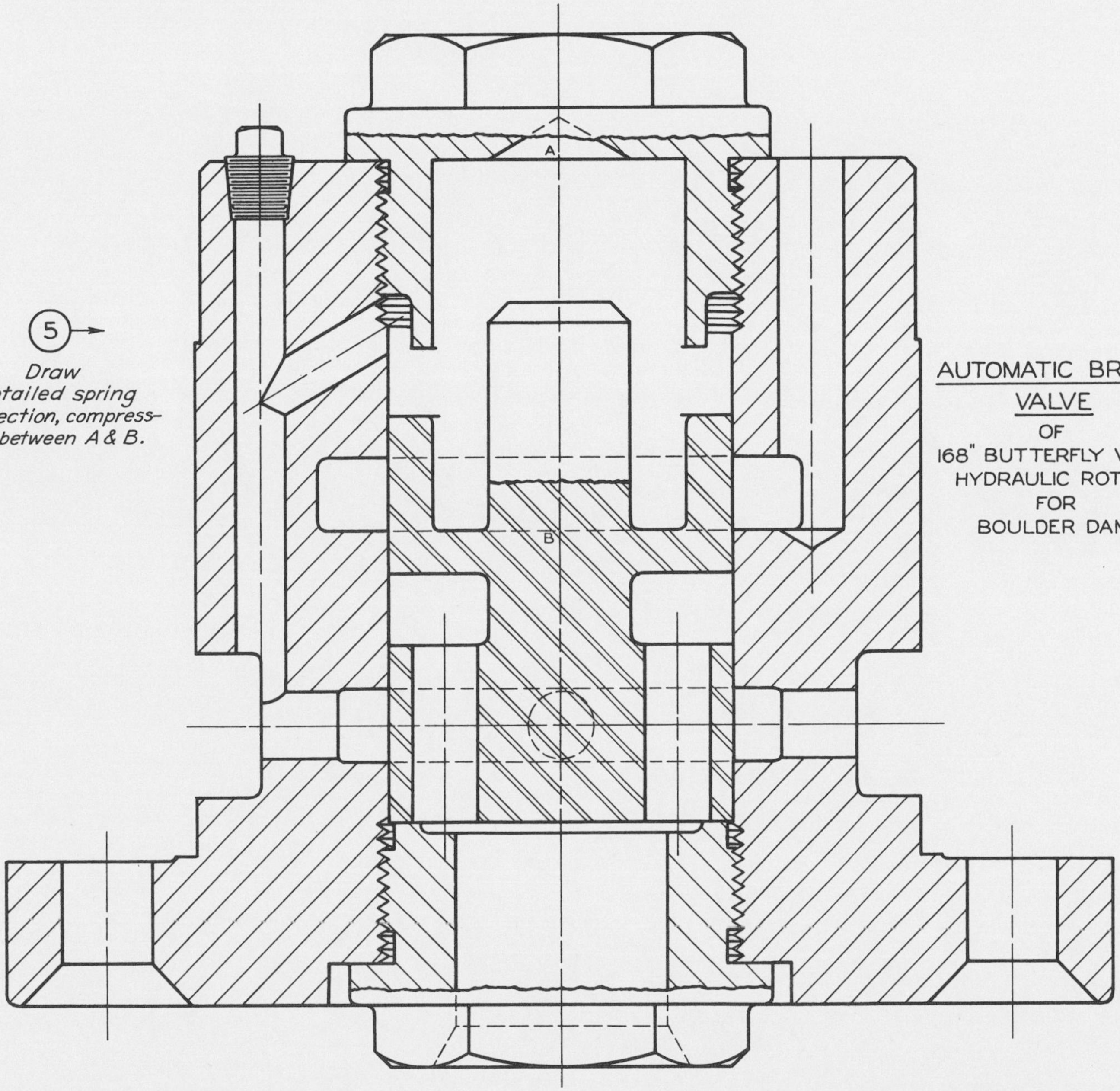
A
B

AUTOMATIC BRAKE VALVE
OF
168" BUTTERFLY VALVE
HYDRAULIC ROTOR
FOR
BOULDER DAM

SECTION	DRAWN BY:		SHEET 92

(1) DRAW PIE CHART.

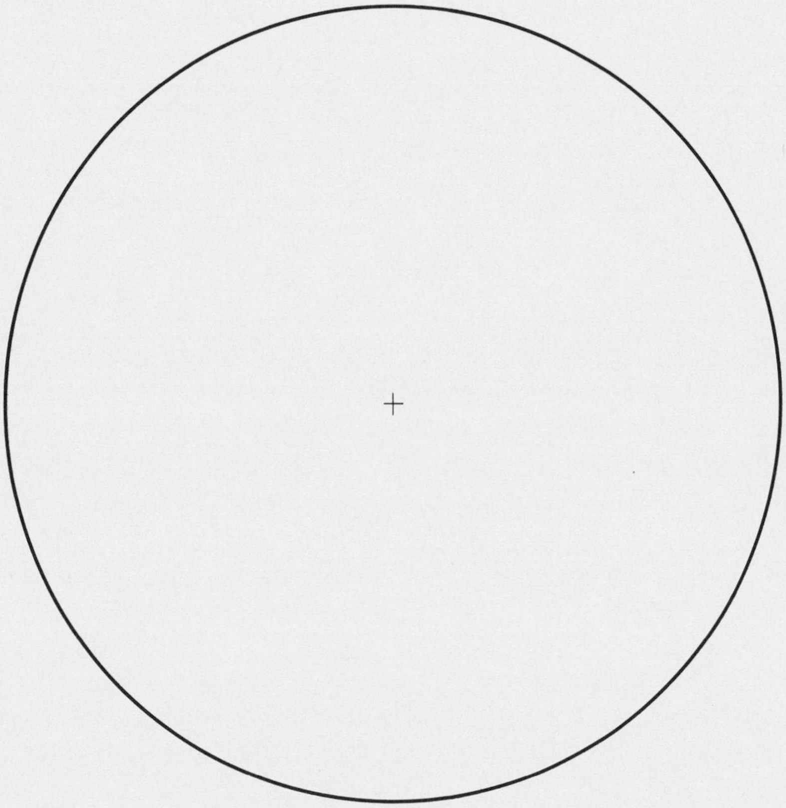

(2) DRAW BAR CHART.

PRODUCTIVITY GAINS

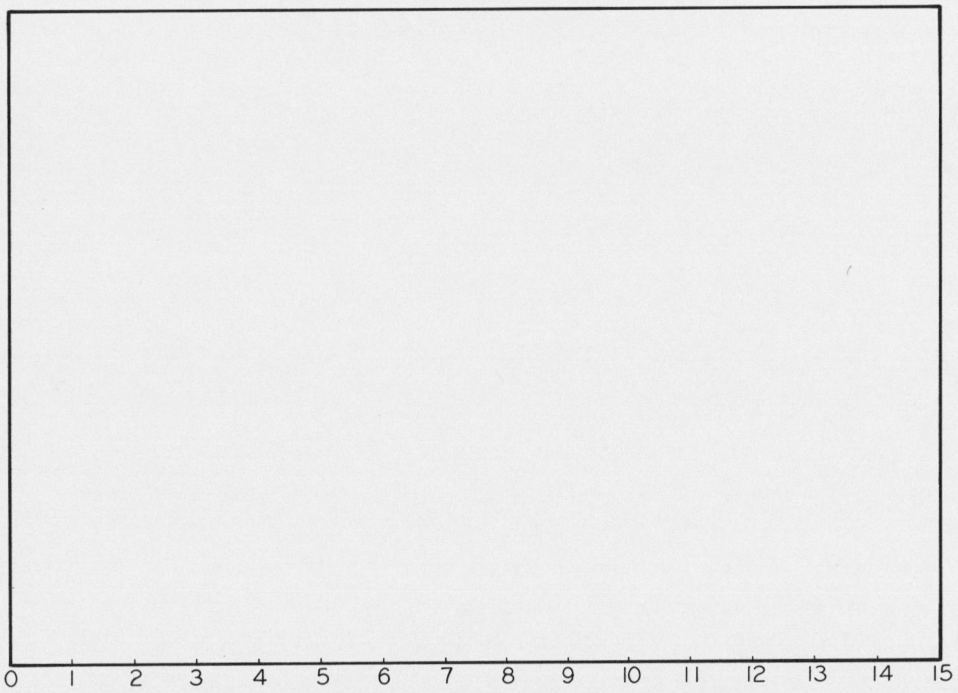
0
1
2
3
4
5
6
7
8
9
10
11
12
13
14
15

ANNUAL PERCENTAGE INCREASE IN OUTPUT PER MAN-HOUR
IN MANUFACTURING
1960 – 1973

SECTION	DRAWN BY:		SHEET 93

TIMBER AND CONCRETE IN COMPRESSION

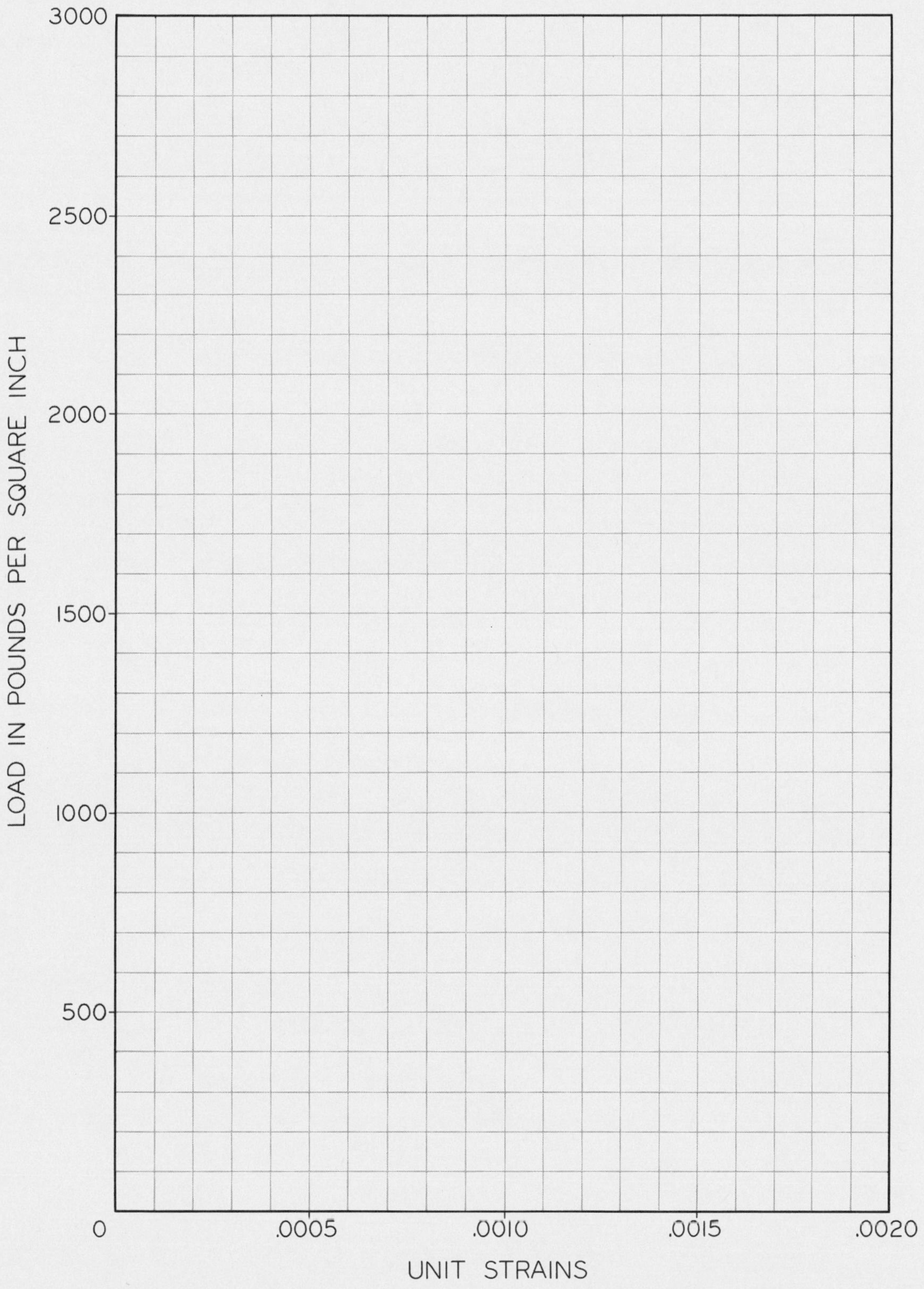
3000
2500
2000
1500
1000
500
0
LOAD IN POUNDS PER SQUARE INCH
.0005
.0010
.0015
.0020
UNIT STRAINS

SECTION	DRAWN BY:		SHEET 94

(1) Find solutions for the equations:

$2y + x = 15$ &

$y - \frac{x}{2} = -4$

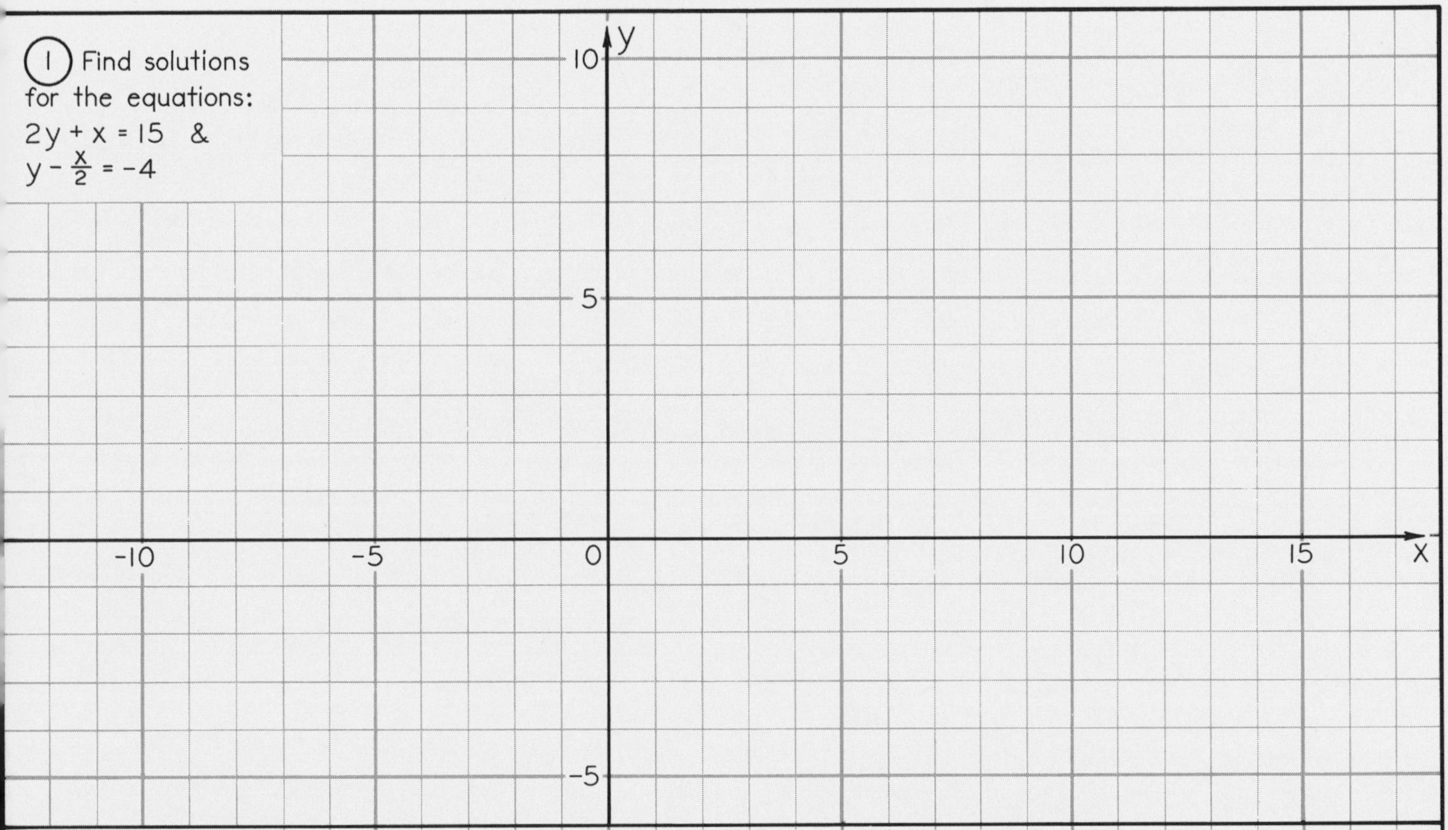
y
10
5
-5
-10
-5
0
5
10
15
x

(2) Find solutions for the equations:

$xy = 20$ &

$2y - x = -8$

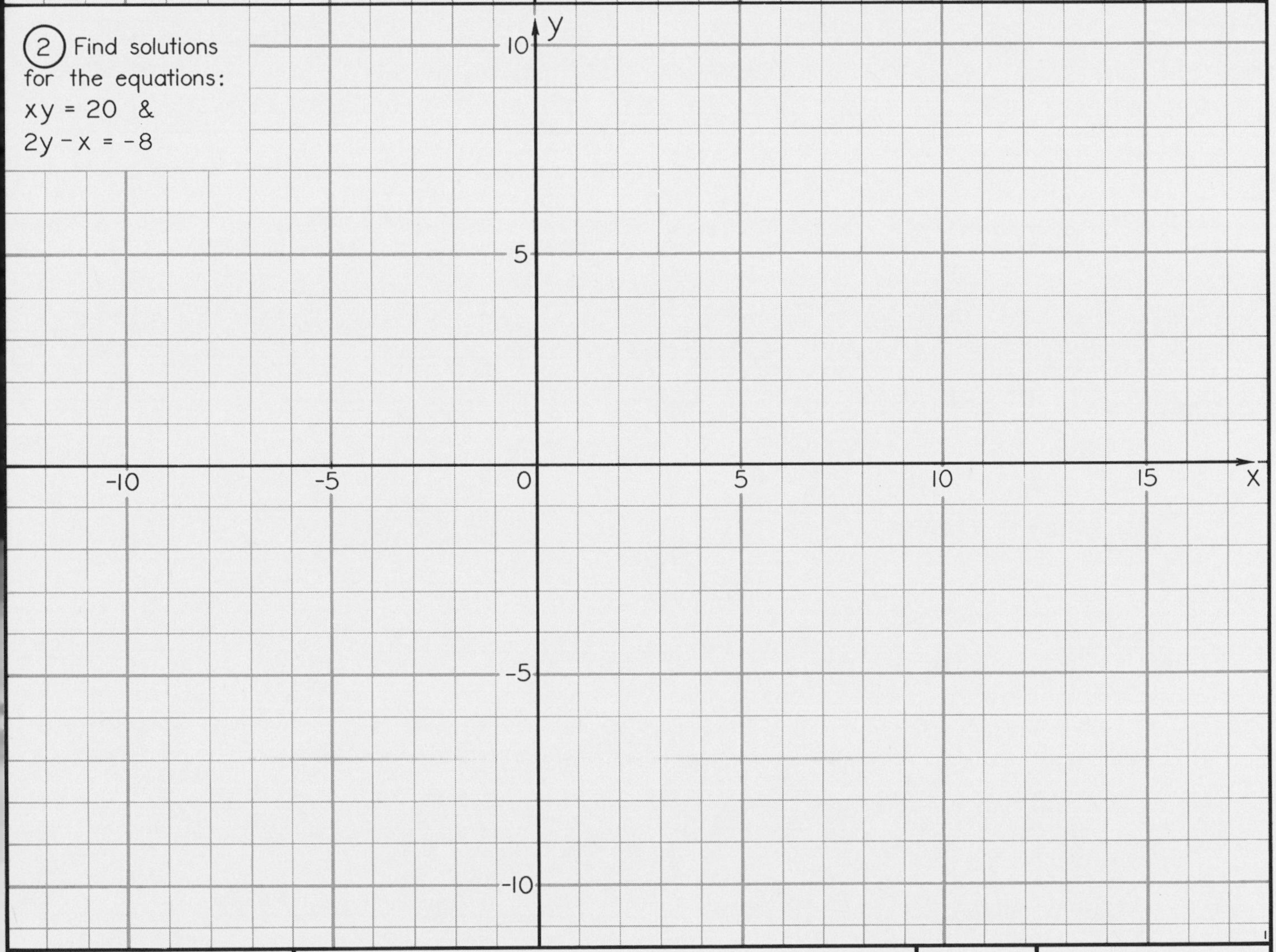
y
10
5
-5
-10
-10
-5
0
5
10
15
x

SECTION	DRAWN BY:		SHEET 95

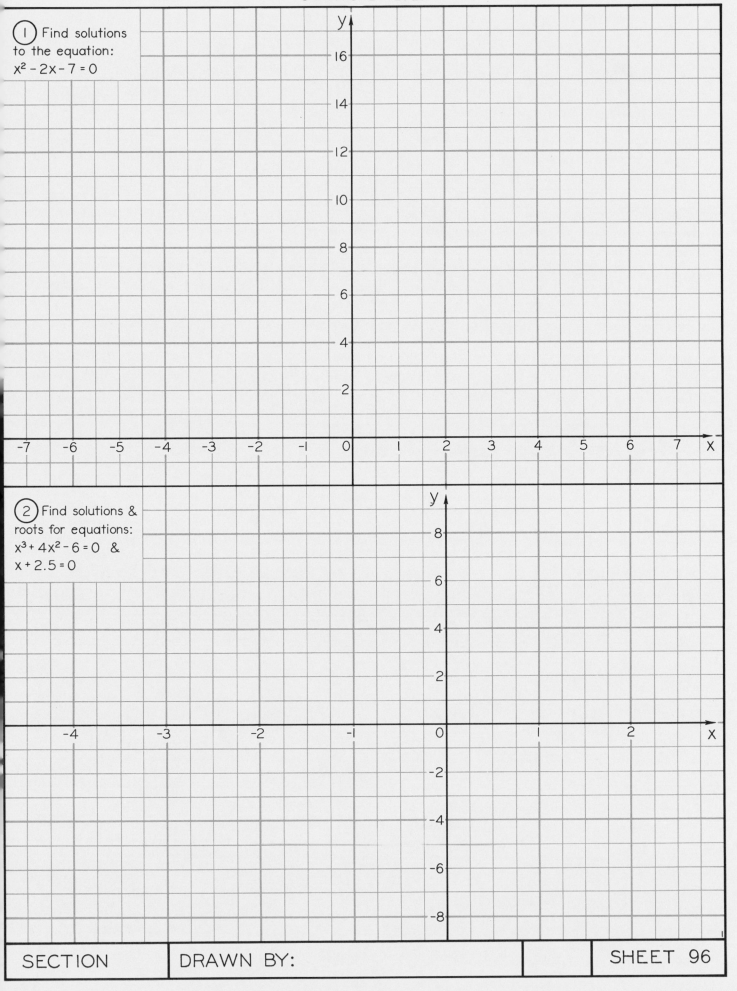

(1) Find solutions to the equation:
$x^2 - 2x - 7 = 0$

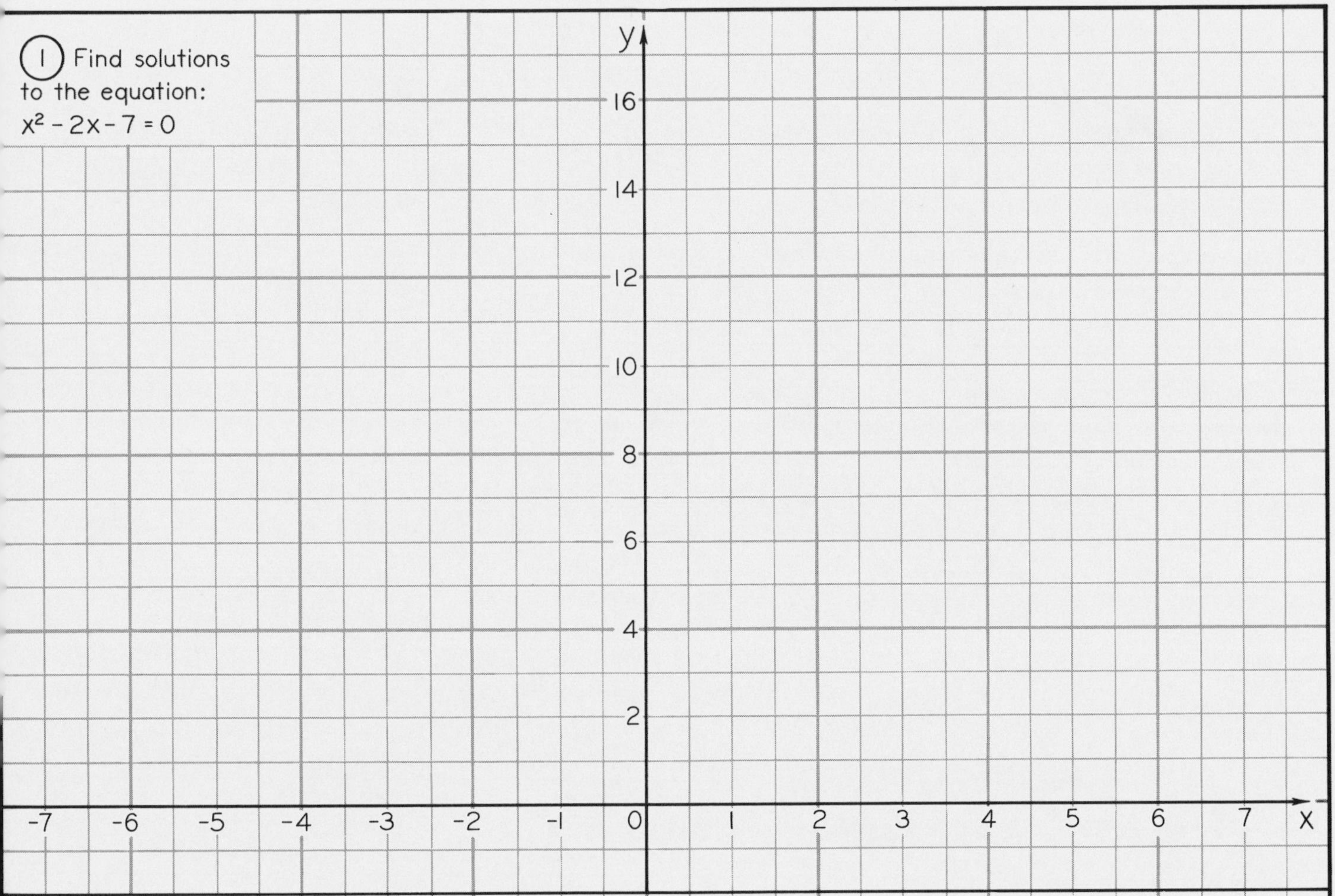
y
16
14
12
10
8
6
4
2
-7 -6 -5 -4 -3 -2 -1 0 1 2 3 4 5 6 7
x

(2) Find solutions & roots for equations:
$x^3 + 4x^2 - 6 = 0$ &
$x + 2.5 = 0$

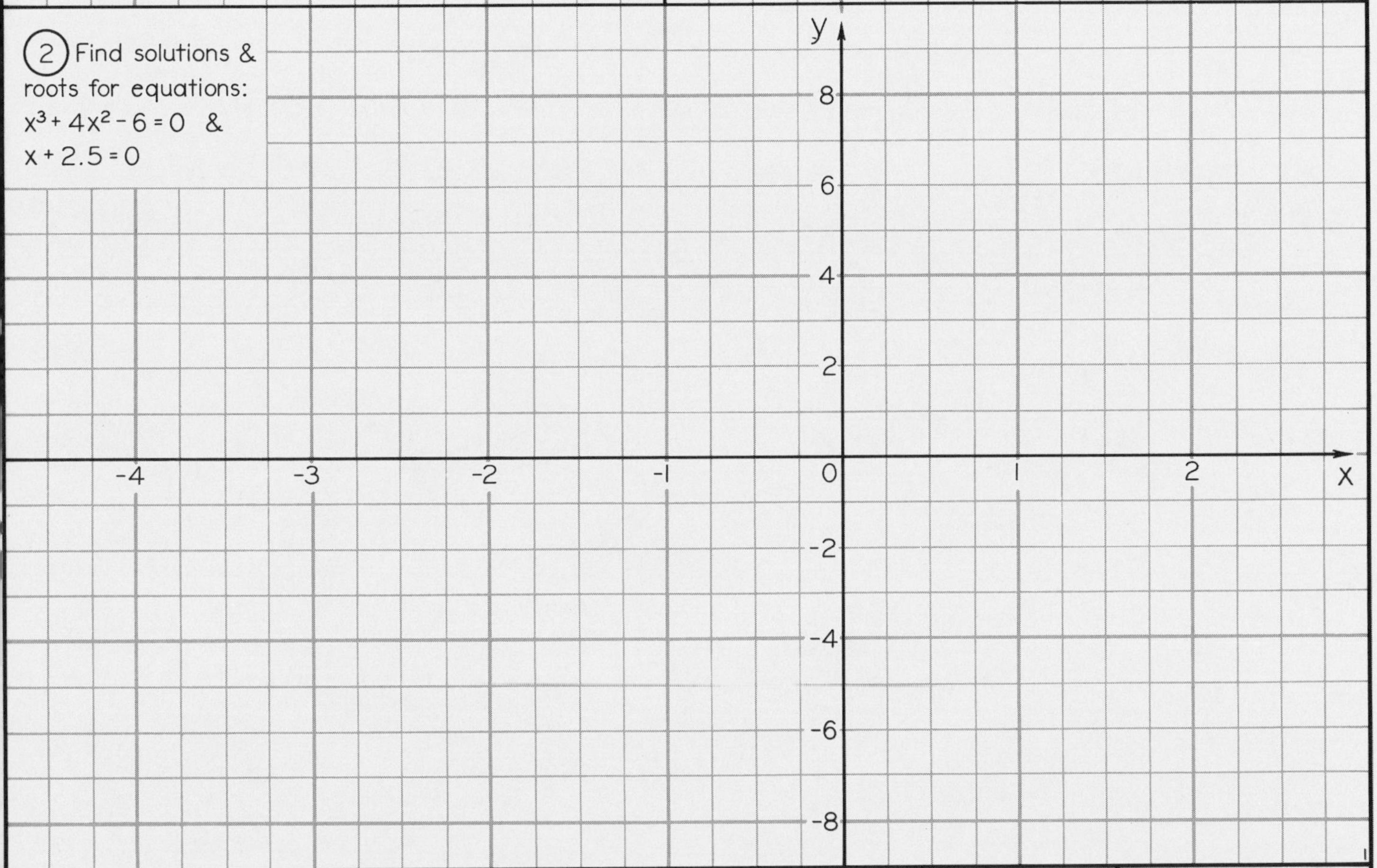
y
8
6
4
2
-4 -3 -2 -1 0 1 2
x
-2
-4
-6
-8

SECTION	DRAWN BY:		SHEET 96

X
1 2 3 4 5 6 7 8 9 10
20 19 18 17 16 15 14 13 12 11 10 9 8 7 6 5 4
y
100 90 80 70 60 50 40 30 20 10 9 8 7 6 5 4 3 2 1
y
0 1 2 3 4 5 6 7 8 9 10
X

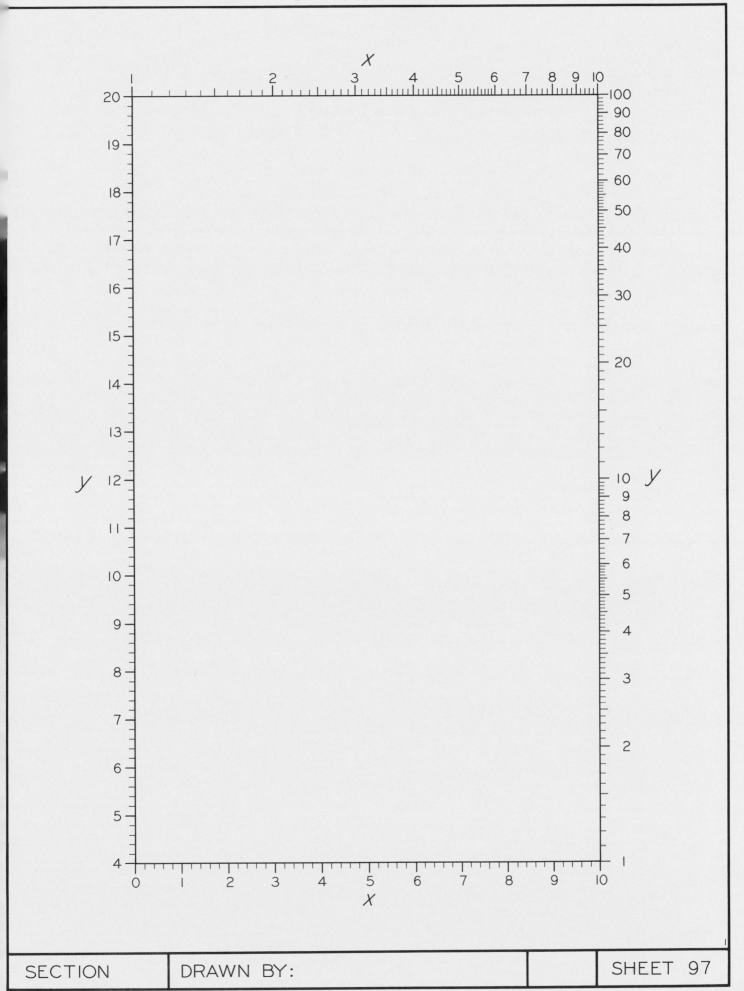

SECTION	DRAWN BY:		SHEET 97

(1)

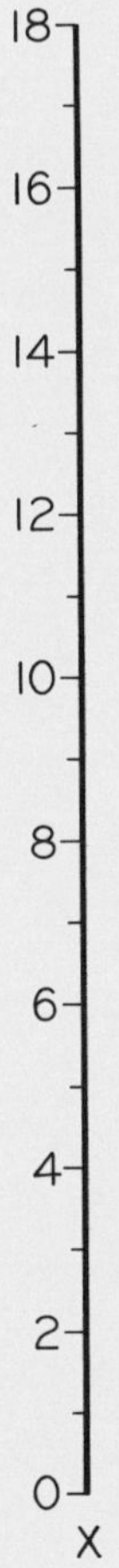
18
16
14
12
10
8
6
4
2
0
X

$X + y - 2 = Z$

(2)

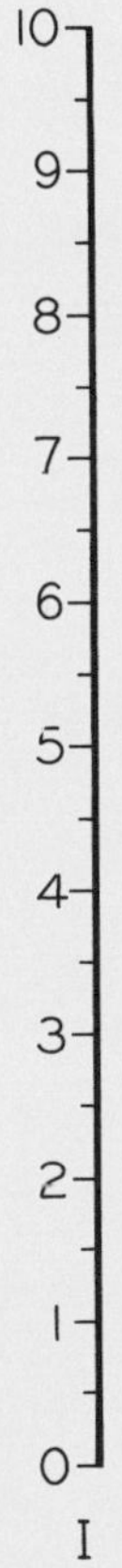
10
9
8
7
6
5
4
3
2
1
0
I

$V = IR$

SECTION	DRAWN BY:		SHEET 98

BOTTOM OF DRAWING

SECTION	DRAWN BY:		SHEET 99

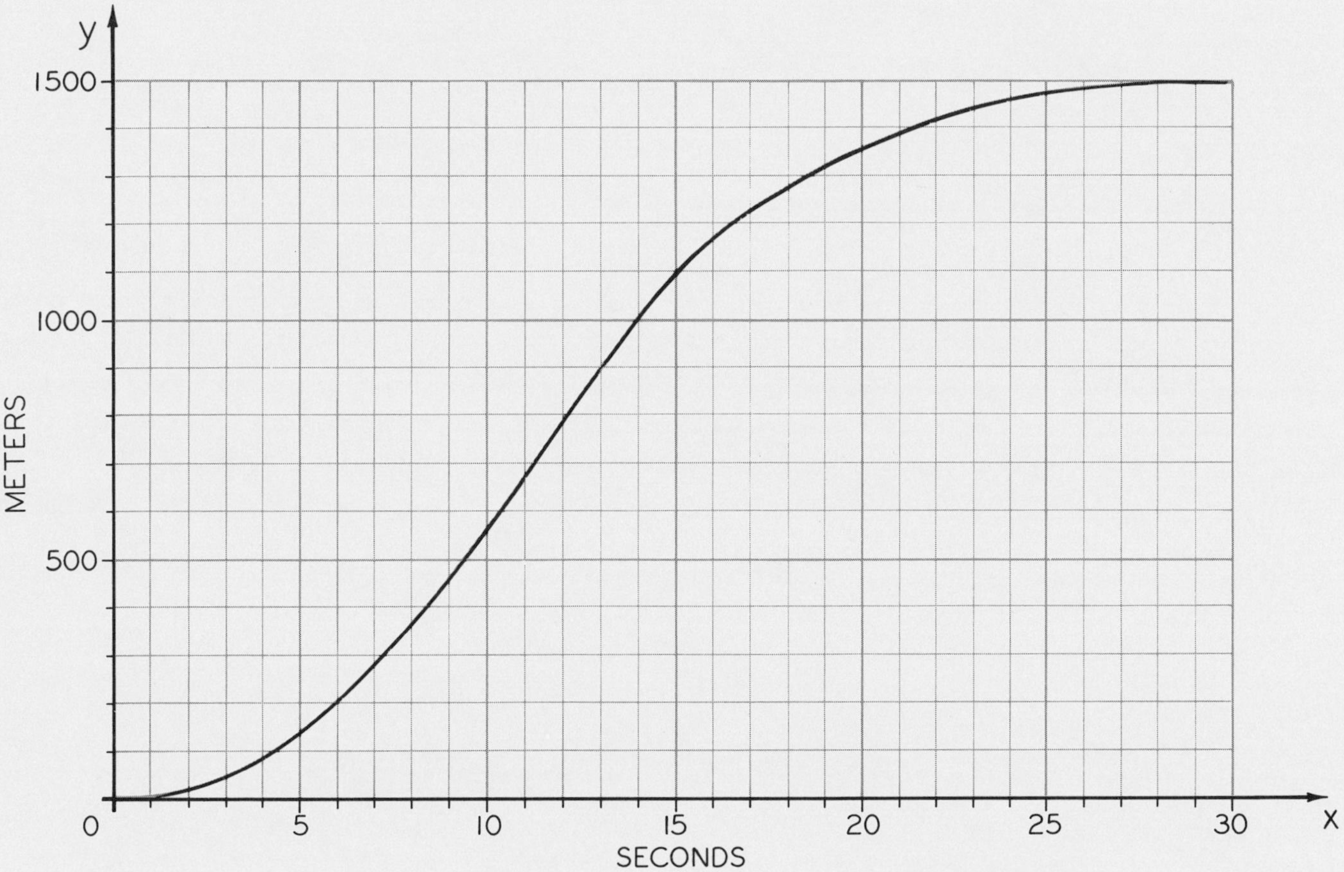
y
1500
1000
500
METERS
0
5
10
15
20
25
30
x
SECONDS

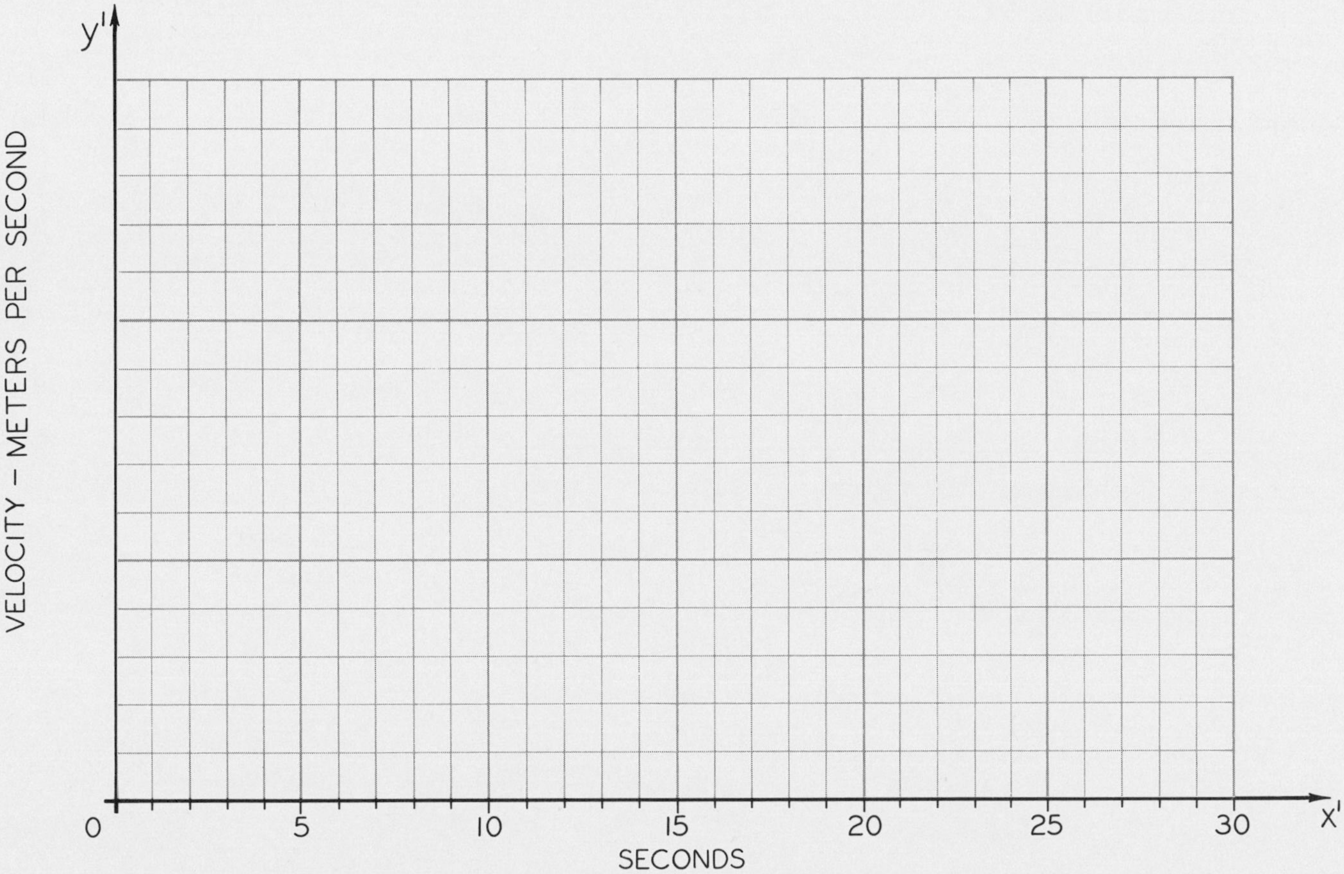
y'
VELOCITY – METERS PER SECOND
0
5
10
15
20
25
30
x'
SECONDS

Velocity after 12 seconds acceleration: ______________________

SECTION	DRAWN BY:		SHEET 100

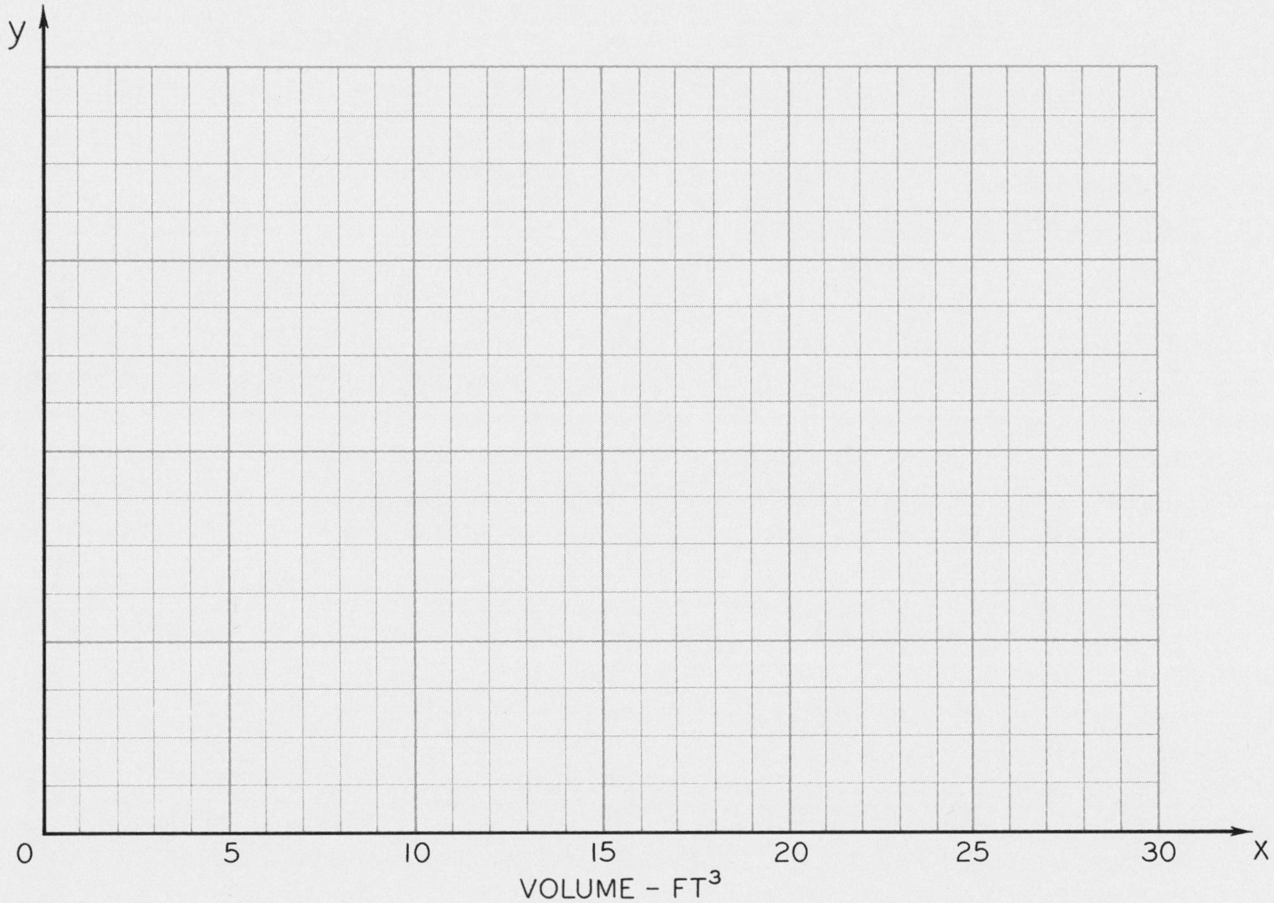
y
0
5
10
15
20
25
30
X
VOLUME - FT³

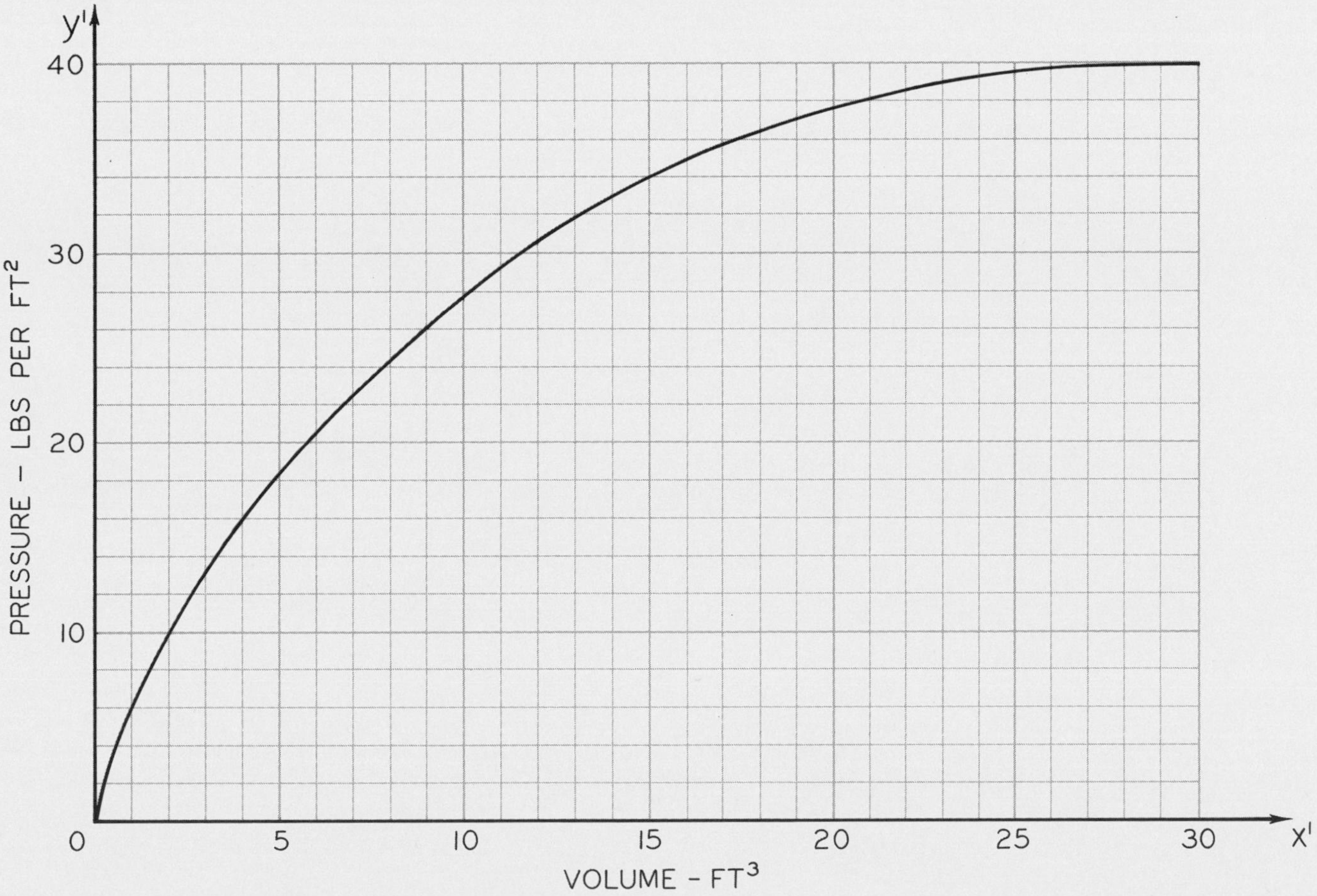
y'
40
30
20
10
0
PRESSURE - LBS PER FT²
5
10
15
20
25
30
X'
VOLUME - FT³

SECTION	DRAWN BY:		SHEET 101

SCALE: ____________

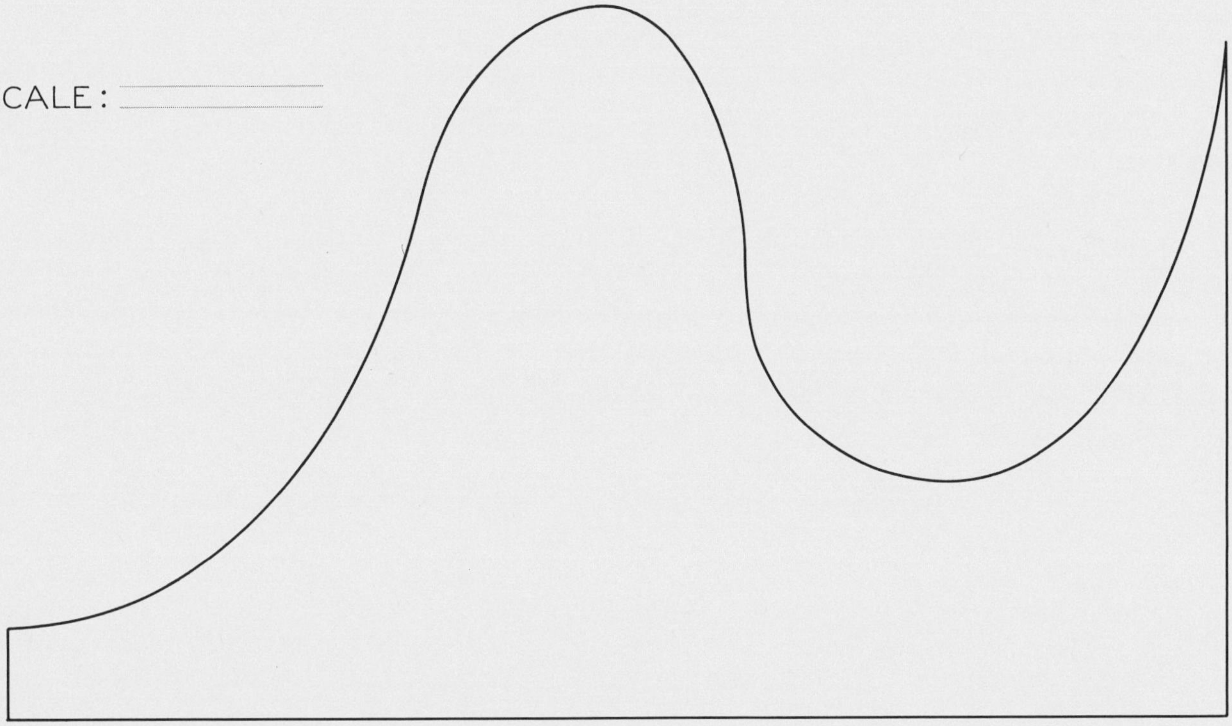

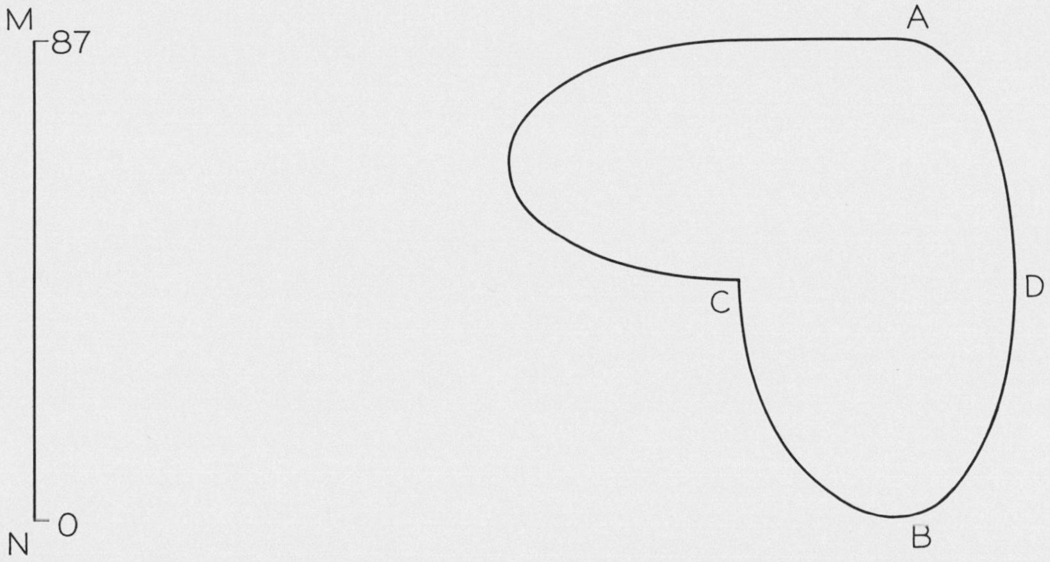
M
87
0
N
A
C
D
B

CROSS SECTIONAL AREA: ____________ TANK LENGTH: ____________

SECTION	DRAWN BY:		SHEET 102

Complete the table of TERMS by entering the letter identifiers of the matching descriptions.

TERMS

TERMS	
CURSOR	
DIGITIZER TABLET	
GRAPHIC PRIMITIVE	
PIXEL	
RESOLUTION	
RASTER DISPLAY	
RAM	
DEBUG	
HARD COPY	
ANALOG	
DIGITAL	
CAE	
COMMAND	
PLOTTER	
HARD DISK	
VECTOR	
COORDINATE SYSTEM	
MENU	
BIT	
MOUSE	
SOFTWARE	
JOY STICK	
WINDOW	
TRANSFORM	
BYTE	
LIGHTPEN	

Descriptions

A Handheld pointing device for pick and coordinate entry

B Computer program to perform specific tasks

C Counts in discrete steps or digits

D Smallest unit of digital information

E Collection of commands for selection

F Device to convert analog picture to coordinate digital data

G Fundamental drawing entity

H Picture element dot in a display grid

I Random Access Memory - volatile physical memory

J Continuous measurements without steps

K Computer assisted engineering

L Group of 8 bits commonly used to represent a character

M Paper printout

N Hand controlled lever used as input device

O Smallest spacing between CRT display elements

P Convert an image into a proper display format

Q Directed line segment with magnitude

R Flicker-free scanned CRT surface

S A bounded rectangular area on screen

T A visual tracking symbol

U Handheld photosensitive input device

V Control signal

W Correct errors

X Non-volatile external storage device

Y Hard copy device for vector drawing

Z Common reference system for spatial relationships

① Complete the table by defining X and Y coordinates of the given points.

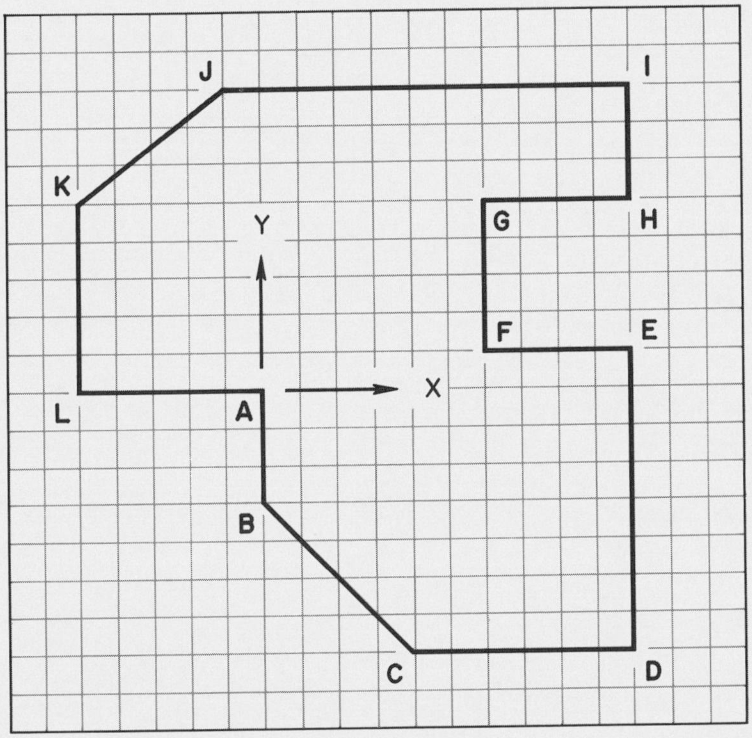
J
I
K
Y
G
H
F
E
X
L
A
B
C
D

Point	Coordinate	
	X	Y
A		
B		
C		
D		
E		
F		
G		
H		
I		
J		
K		
L		

② Plot the given points on the grid and draw the view.

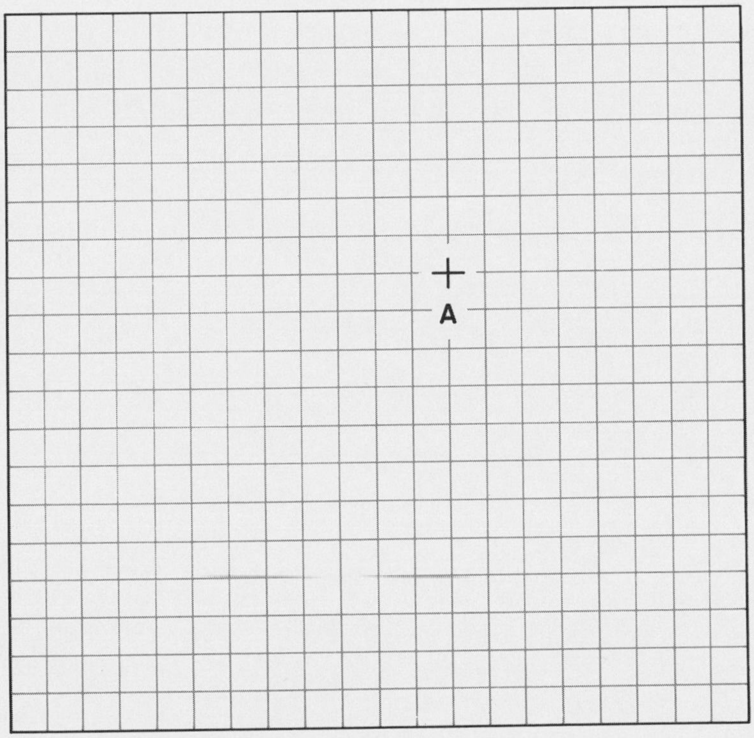
A

Point	Coordinate	
	X	Y
A	0	0
B	0	4
C	-5	4
D	-5	-2
E	-10	-6
F	-10	-10
G	-2	-10
H	6	-4
I	6	3
J	3	3
K	3	-2
L	0	-4

SECTION	DRAWN BY:		SHEET 104

CAD: THREE DIMENSIONAL COORDINATE PLOT

① Complete the table for drawing the object.

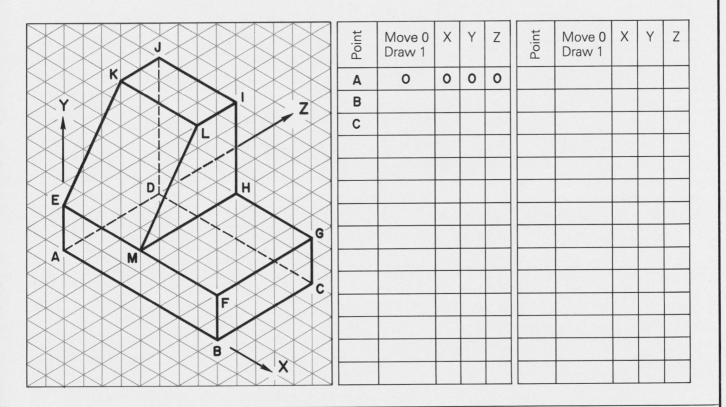

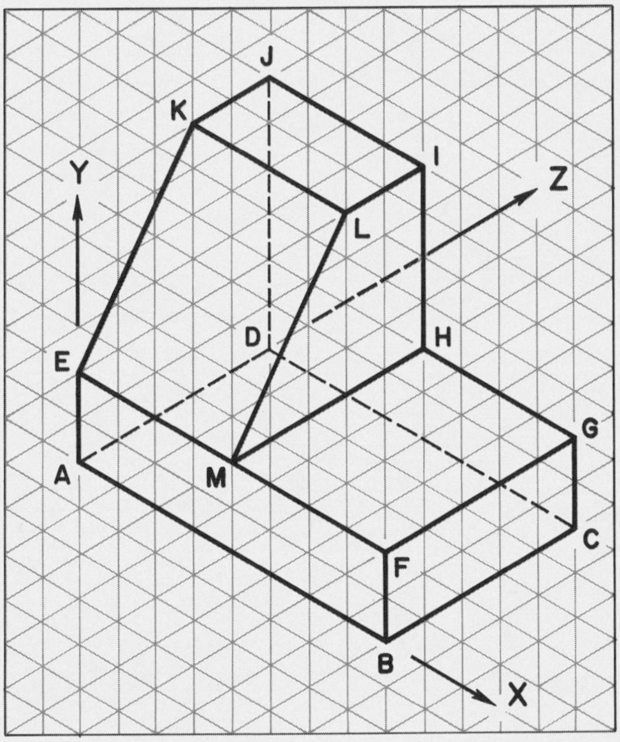

J
K
I
Y
Z
L
D
H
E
G
A
M
C
F
B
X

Point	Move 0 Draw 1	X	Y	Z
A	0	0	0	0
B				
C				

Point	Move 0 Draw 1	X	Y	Z

② Draw the object based on the data given in the table.

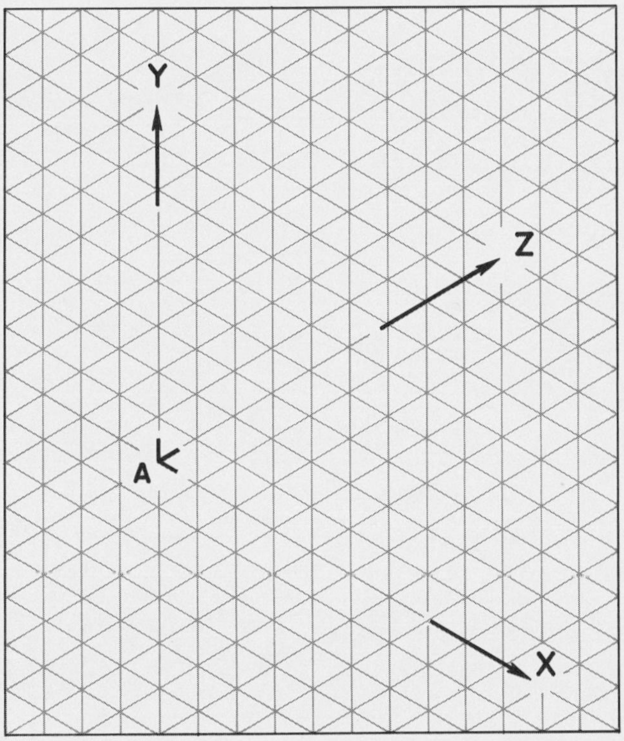

Y
Z
A
X

Point	Move 0 Draw 1	X	Y	Z
A	0	0	0	0
B	1	0	5	0
C	1	0	5	5
D	1	0	0	5
E	1	6	0	5
F	1	6	0	0
A	1	0	0	0
D	1	0	0	5
G	0	-3	-4	0
L	1	-3	-1	0
K	1	-3	-1	5
J	1	-3	-4	5
I	1	0	-4	5
H	1	0	-4	0

Point	Move 0 Draw 1	X	Y	Z
G	1	-3	-4	0
J	1	-3	-4	5
L	0	-3	-1	0
B	1	0	5	0
K	0	-3	-1	5
C	1	0	5	5
H	0	0	-4	0
F	1	6	0	0
I	0	0	-4	5
E	1	6	0	5

Complete the table by entering the Menu Selections used for generating the drawing.

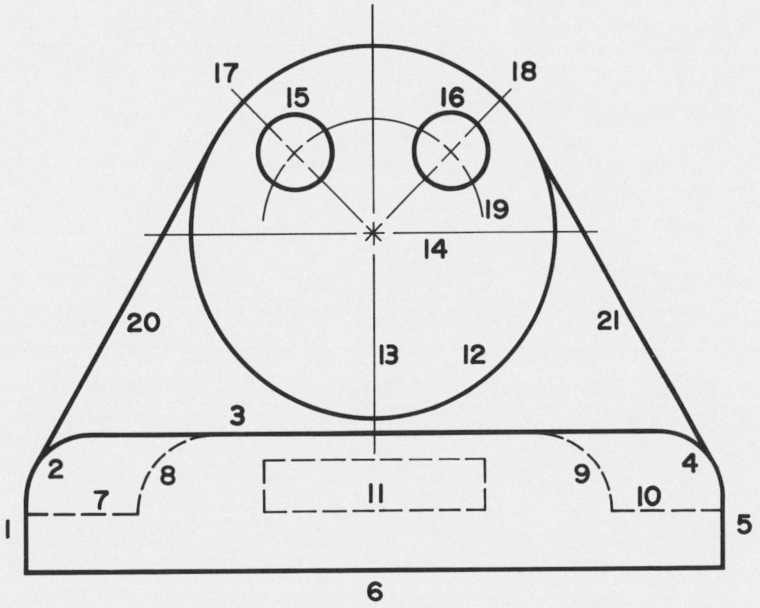
17
15
16
18
19
14
20
21
13
12
3
2
4
7
8
11
9
10
1
5
6

LINE TYPE MENU

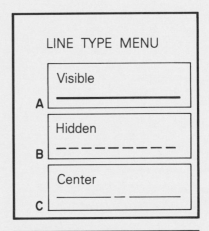

- A — Visible
- B — Hidden
- C — Center

ENTITY MENU

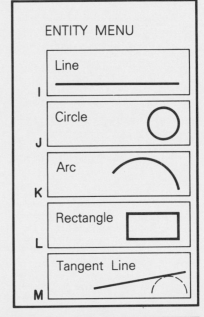

- I — Line
- J — Circle
- K — Arc
- L — Rectangle
- M — Tangent Line

CONSTRUCTION MENU

- P — From point to point
- Q — Around center with radius
- R — Around center with radius and angle
- S — With height and width
- T — From circle to arc

Entity	Line type menu selection	Entity menu selection	Construction menu selection
1			
2			
3			
4			
5			
6			
7			
8			
9			
10			
11			
12			
13			
14			
15			
16			
17			
18			
19			
20			
21			

Description of VIEW COORDINATES

VIEW COORDINATES are the coordinate values of the object as assigned with respect to the computer screen, with X , Y and Z axes positioned as shown below . The coordinates remain the same irrespective of the view selected on the screen .

Axis	Position	Positive Direction
X	Horizontal	To the right
Y	Vertical	Toward the top
Z	Perpendicular to the screen	Outward from the screen

Description of WORLD COORDINATES

WORLD COORDINATES are the coordinate values of the object as assigned with respect to the axes of the object . The X , Y and Z axes are positioned as shown, such that for the top view the X axis is horizontal to the right, the Y axis is vertical to the top and the Z axis is perpendicular to the screen positioned outwards . The coordinates in relation to the screen change according to the view selected on the screen.

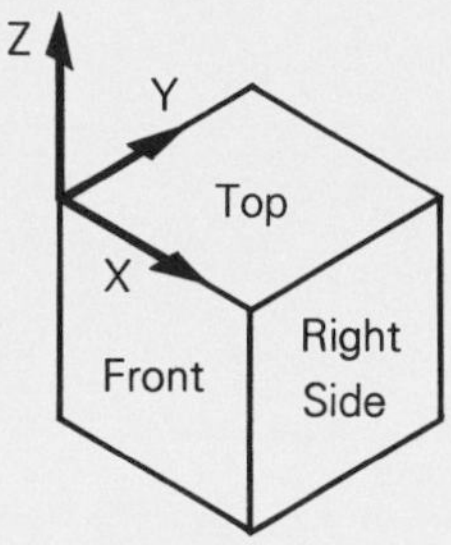
Z
Y
X
Top
Front
Right
Side

Complete the tables by entering the VIEW and WORLD COORDINATES of the given points of the object .

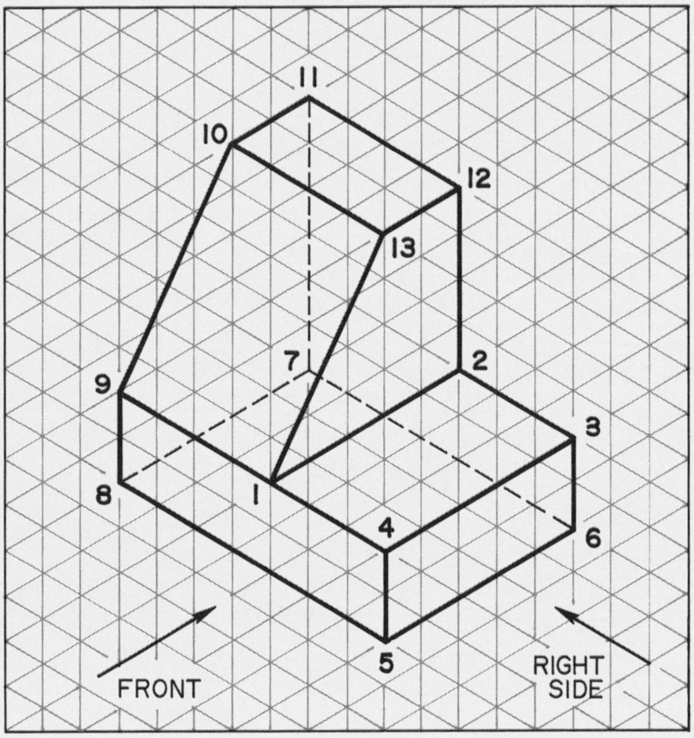
11
10
12
13
7
2
9
3
8
1
4
6
5
FRONT
RIGHT
SIDE

FRONT VIEW						
Points	View Coordinates			World Coordinates		
	X	Y	Z	X	Y	Z
1						
2						
3						
4						
5						
6						
7						
8						
9						
10						
11						
12						
13						

RIGHT SIDE VIEW						
Points	View Coordinates			World Coordinates		
	X	Y	Z	X	Y	Z
1						
2						
3						
4						
5						
6						
7						
8						
9						
10						
11						
12						
13						

1 SPACER ARM
C1—2 REQD

METRIC

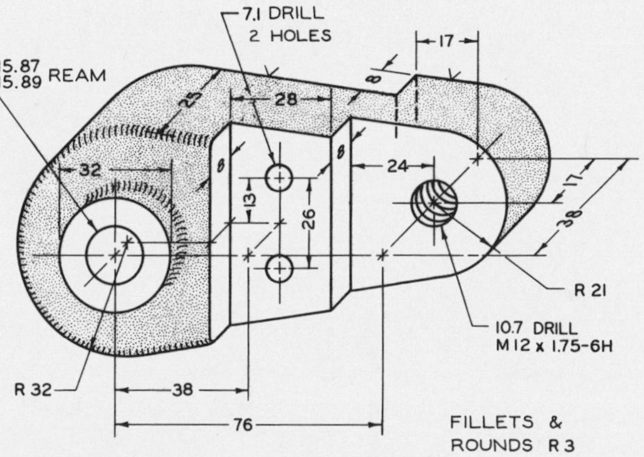
7.1 DRILL
2 HOLES
15.87
15.89 REAM
17
28
8
25
32
13
26
24
17
38
R 21
10.7 DRILL
M12 x 1.75-6H
R 32
38
76

FILLETS & ROUNDS R 3

2 DRIVE TIGHTENER BASE
C1—1 REQD

FILLETS & ROUNDS .06 R

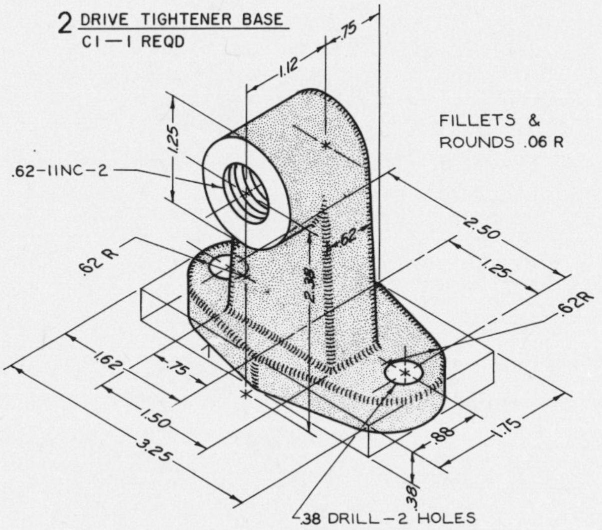
.75
1.12
1.25
.62-11NC-2
.62
.62 R
2.38
2.50
1.25
.62R
1.62
.75
1.50
3.25
.88
1.75
.38
.38 DRILL—2 HOLES

3 BASE ANGLE FLANGE
C1—1 REQD

METRIC

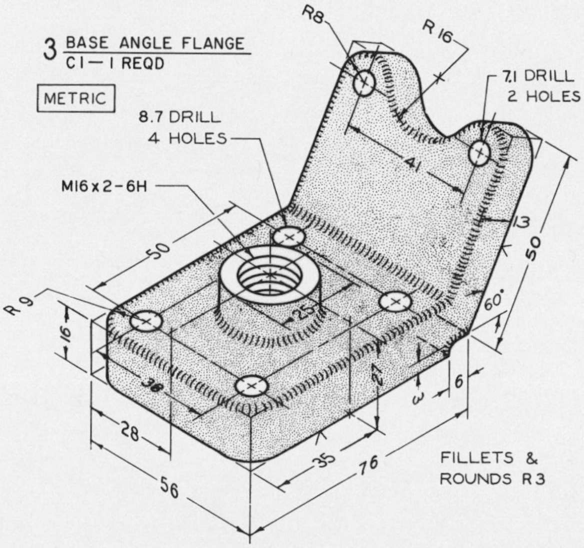
R8
R 16
7.1 DRILL
2 HOLES
8.7 DRILL
4 HOLES
M16 x 2-6H
41
13
50
50
R 9
16
25
60°
38
27
3
6
28
35
76
56

FILLETS & ROUNDS R3

4 CENTER GUIDE
CRS—2 REQD

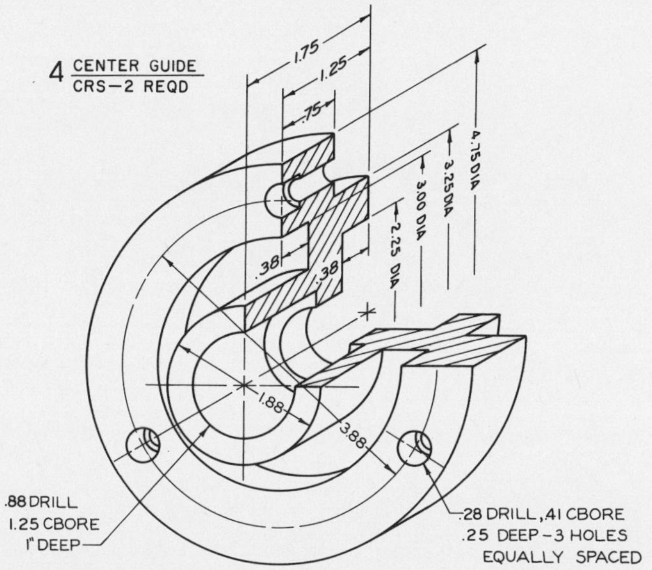
1.75
1.25
.75
4.75 DIA
3.25 DIA
3.00 DIA
2.25 DIA
.38
.38
1.88
3.88
.88 DRILL
1.25 CBORE
1" DEEP
.28 DRILL, .41 CBORE
.25 DEEP—3 HOLES
EQUALLY SPACED

5 BELT TAKE-UP ARM
C1—1 REQD

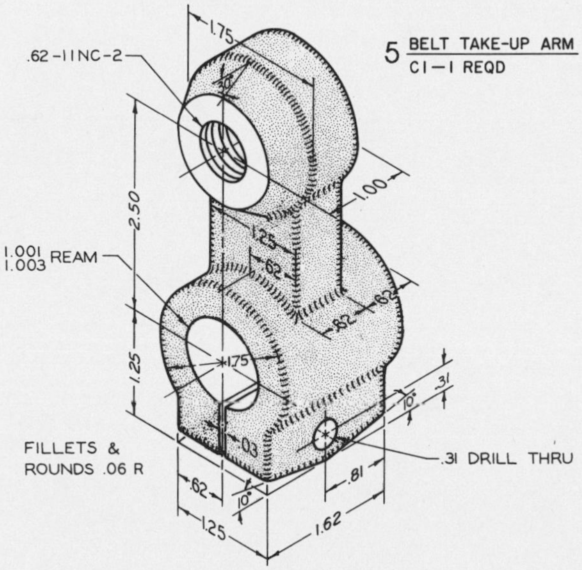
.62-11NC-2
1.75
.50
1.00
2.50
1.001
1.003 REAM
1.25
.62
.62
.62
1.75
1.25
.31
10°
.03
.31 DRILL THRU
.62
.81
10°
1.25
1.62

FILLETS & ROUNDS .06 R

6 TRIP BRACKET
C1—1 REQD

METRIC

FILLETS AND ROUNDS R 3

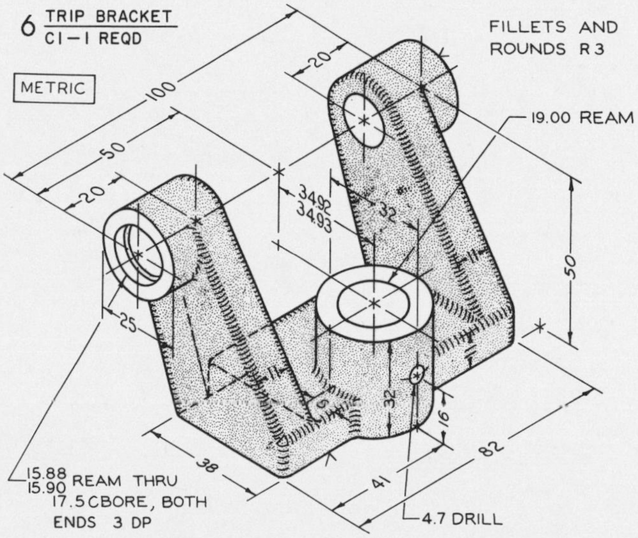
20
100
50
20
19.00 REAM
34.92
34.93
32
50
25
6
32
16
38
41
82
15.88
15.90 REAM THRU
17.5 CBORE, BOTH
ENDS 3 DP
4.7 DRILL

DETAIL DRAWINGS	DRAW OR SKETCH THE NECESSARY VIEWS OF THE OBJECT ASSIGNED. DIMENSION COMPLETELY	SHEET 109

SECTION	DRAWN BY:		SHEET

SECTION	DRAWN BY:		SHEET

SECTION	DRAWN BY:		SHEET

SECTION	DRAWN BY:		SHEET

SECTION	DRAWN BY:		SHEET

SECTION	DRAWN BY:		SHEET

SECTION	DRAWN BY:		SHEET

(1)

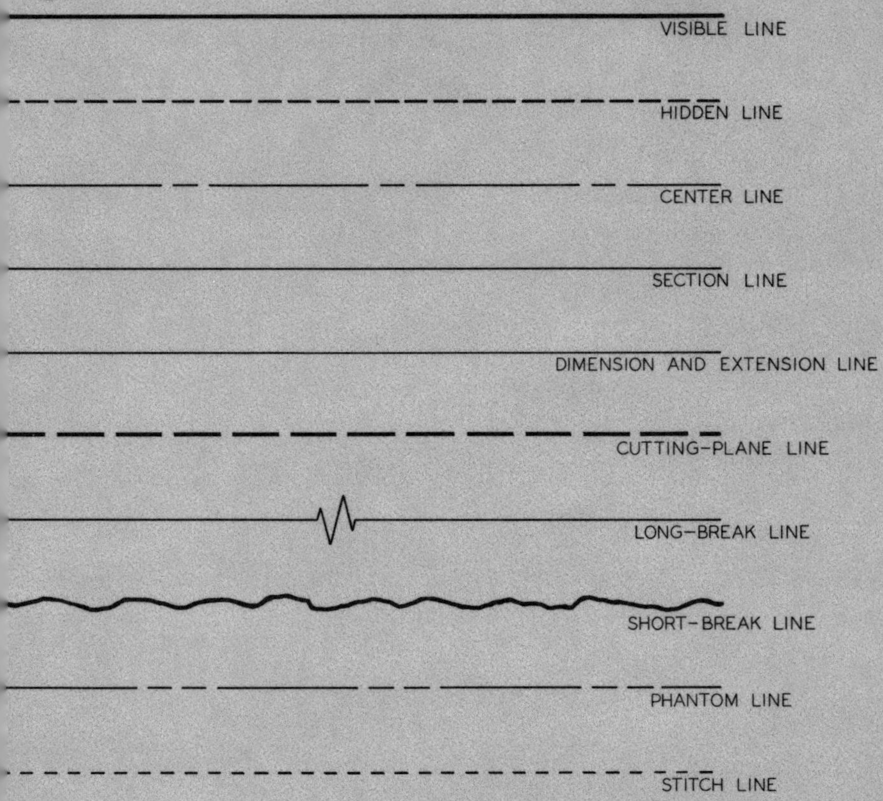
VISIBLE LINE
HIDDEN LINE
CENTER LINE
SECTION LINE
DIMENSION AND EXTENSION LINE
CUTTING-PLANE LINE
LONG-BREAK LINE
SHORT-BREAK LINE
PHANTOM LINE
STITCH LINE

(2)

A

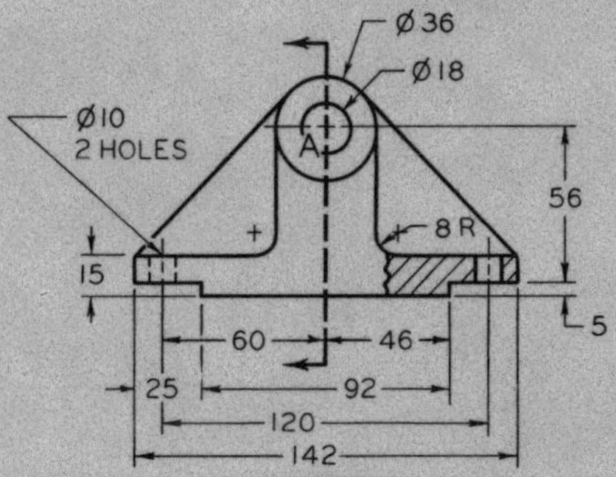
Ø36
Ø18
Ø10
2 HOLES
A
56
8 R
15
5
60
46
25
92
120
142

METRIC

SECTION	DRAWN BY:		SHEET 11

(1)

VISIBLE LINE

HIDDEN LINE

CENTER LINE

SECTION LINE

DIMENSION AND EXTENSION LINE

CUTTING-PLANE LINE

LONG-BREAK LINE

SHORT-BREAK LINE

PHANTOM LINE

STITCH LINE

(2)

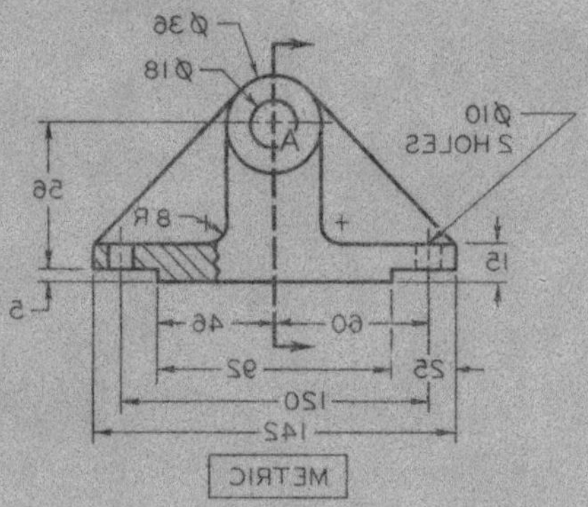
Ø36
Ø18
Ø10
2 HOLES
A
56
8 R
15
5
46
60
92
25
120
142
METRIC

+A

(1) Draw horizontal lines to scales or sizes indicated. Draw cross-bars at ends of lines.

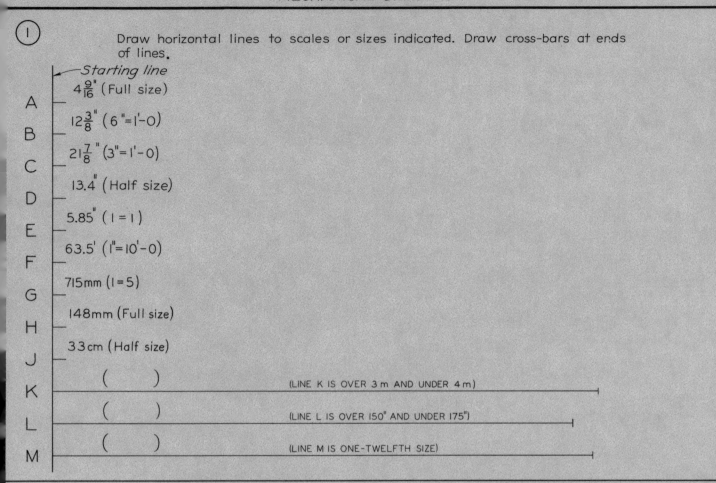

Starting line

A — $4\frac{9}{16}$" (Full size)

B — $12\frac{3}{8}$" (6"=1'-0)

C — $21\frac{7}{8}$" (3"=1'-0)

D — 13.4" (Half size)

E — 5.85" (1 = 1)

F — 63.5' (1"=10'-0)

G — 715mm (1=5)

H — 148mm (Full size)

J — 33cm (Half size)

K — (   ) (LINE K IS OVER 3 m AND UNDER 4 m)

L — (   ) (LINE L IS OVER 150" AND UNDER 175")

M — (   ) (LINE M IS ONE-TWELFTH SIZE)

(2)

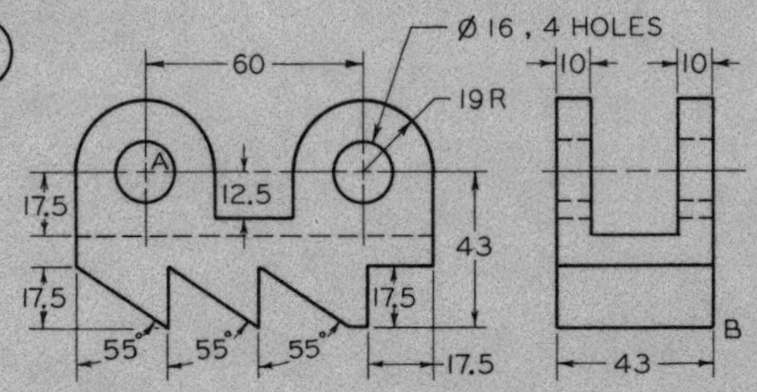
Ø16, 4 HOLES
60
19R
10
10
A
12.5
17.5
43
17.5
17.5
55°
55°
55°
17.5
43
B

MAKE INSTRUMENTAL PENCIL DRAWING, LOCATING VIEWS AT A AND B. SCALE: 1 = 1 METRIC

A

B

SECTION	DRAWN BY:		SHEET 15

(1) Draw horizontal lines to scales or sizes indicated. Draw cross-bars at ends of lines.

Starting line

A	4 9/16" (Full size)
B	12 3/8" (6"=1'-0)
C	21 7/8" (3"=1'-0)
D	13.4" (Half size)
E	5.85" (1 = 1)
F	63.5' (1"=10'-0)
G	715mm (1=5)
H	148mm (Full size)
J	33cm (Half size)
K	( ) (LINE K IS OVER 3 m AND UNDER 4 m)
L	( ) (LINE L IS OVER 150" AND UNDER 175")
M	( ) (LINE M IS ONE-TWELFTH SIZE)

(2) MAKE INSTRUMENTAL PENCIL DRAWING, LOCATING VIEWS AT A AND B. SCALE: 1 = 1 METRIC

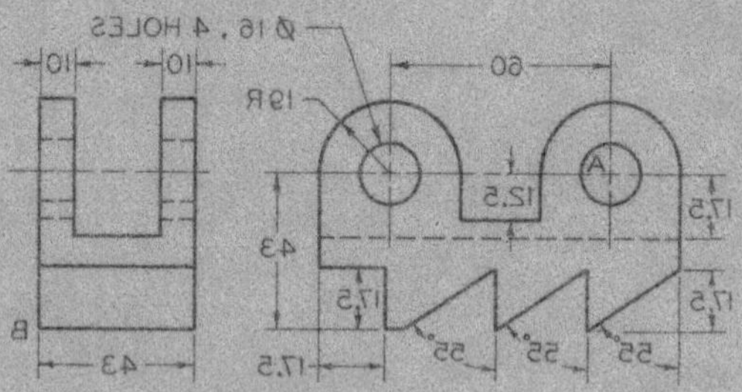

Ø16, 4 HOLES
19R
60
10
10
12.5
17.5
17.5
17.5
43
55
55
55
17.5
43
A
B

A

B

SECTION	DRAWN BY:		SHEET 15

(1) Draw horizontal lines to scales or sizes indicated. Draw cross-bars at ends of lines.

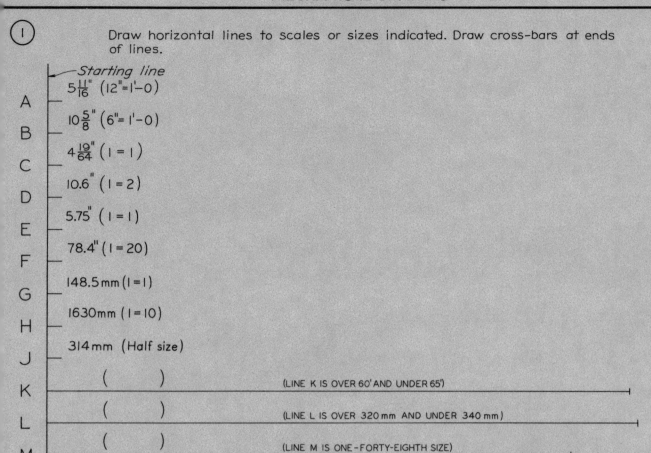

Starting line

A — 5 11/16" (12"=1'-0)

B — 10 5/8" (6"= 1'-0)

C — 4 19/64" (1 = 1)

D — 10.6" (1 = 2)

E — 5.75" (1 = 1)

F — 78.4" (1 = 20)

G — 148.5 mm (1 = 1)

H — 1630 mm (1 = 10)

J — 314 mm (Half size)

K — (   ) (LINE K IS OVER 60' AND UNDER 65')

L — (   ) (LINE L IS OVER 320 mm AND UNDER 340 mm)

M — (   ) (LINE M IS ONE-FORTY-EIGHTH SIZE)

(2)

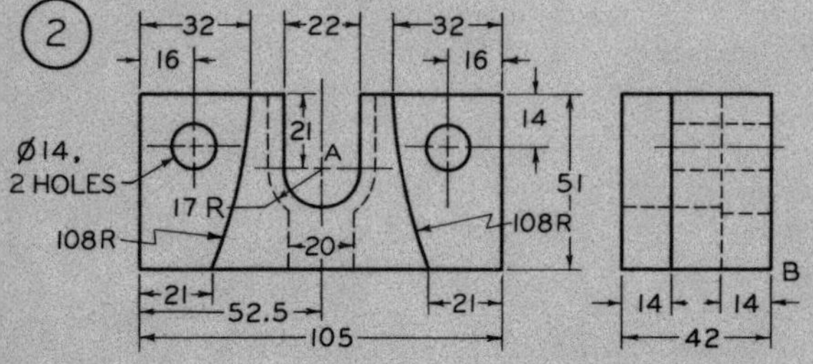
32
22
32
16
16
14
21
A
51
Ø14.
2 HOLES
17 R
108R
108R
20
21
52.5
21
105
14
14
42
B

MAKE INSTRUMENTAL PENCIL DRAWING, LOCATING VIEWS AT A AND B. SCALE: FULL SIZE

METRIC

+A

|B

SECTION	DRAWN BY:		SHEET 16

(1) Draw horizontal lines to scales or sizes indicated. Draw cross-bars at ends of lines.

Starting line

A $5\frac{11}{16}$" (12"=1'-0)

B $10\frac{5}{8}$" (6"=1'-0)

C $4\frac{19}{64}$" (1 = 1)

D 10.6" (1 = 2)

E 5.75" (1 = 1)

F 78.4" (1 = 20)

G 148.5mm (1=1)

H 1630mm (1 = 10)

J 314mm (Half size)

K (   ) (LINE K IS OVER 60' AND UNDER 65')

L (   ) (LINE L IS OVER 320mm AND UNDER 340mm)

M (   ) (LINE M IS ONE-FORTY-EIGHTH SIZE)

(2) MAKE INSTRUMENTAL PENCIL DRAWING, LOCATING VIEWS AT A AND B. SCALE: FULL SIZE

METRIC

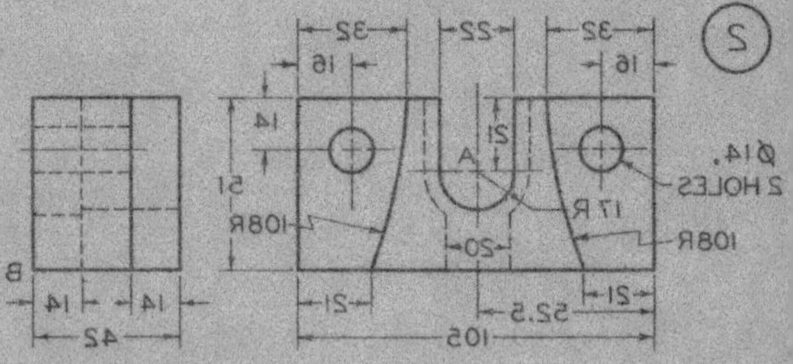
32
22
32
16
16
14
21
51
108R
17 R
Ø14, 2 HOLES
108R
20
21
52.5
21
105
14
14
42
B
A

A

B

SECTION	DRAWN BY:		SHEET 16

GEAR-SHIFTER BLOCK

MATL: CI   SCALE: $\frac{1}{2}$ SIZE

Draw front and top views

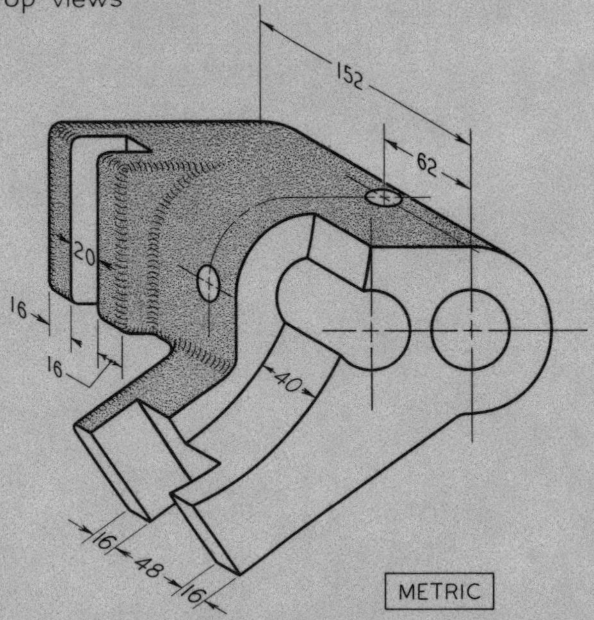
152
62
20
16
16
40
16
48
16

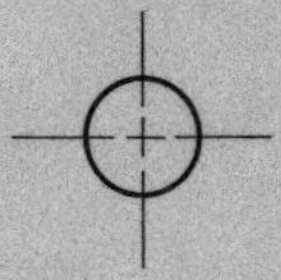

METRIC

FILLETS & ROUNDS 6R

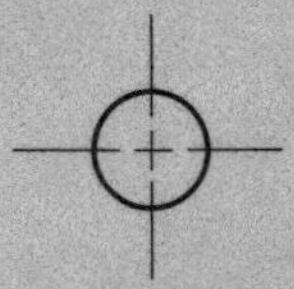

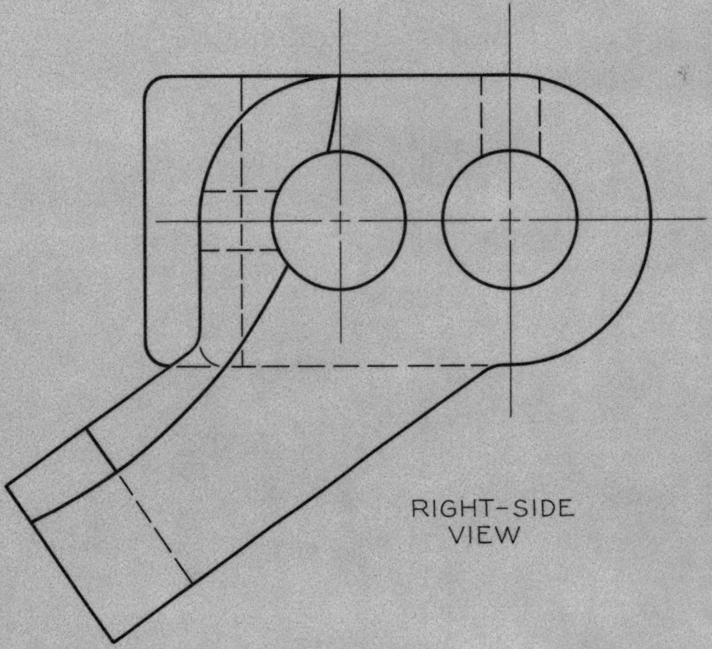

RIGHT-SIDE VIEW

GEAR-SHIFTER BLOCK

MATL: CI SCALE: $\frac{1}{2}$ SIZE

Draw front and top views

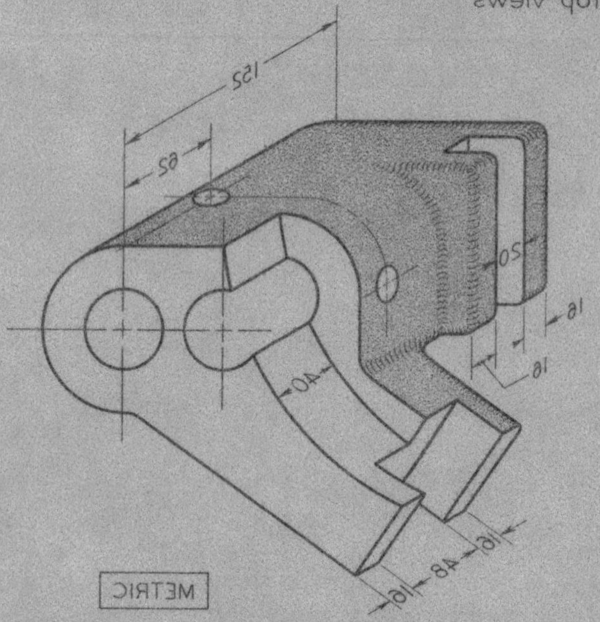
152
62
20
16
16
40
16
48
16

METRIC

FILLETS & ROUNDS 6R

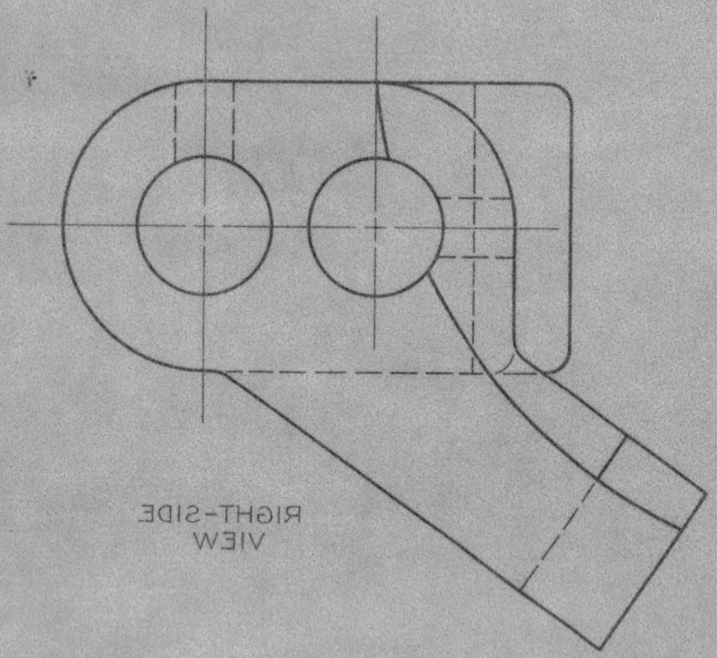

RIGHT-SIDE VIEW

SECTION	DRAWN BY:			SHEET 37

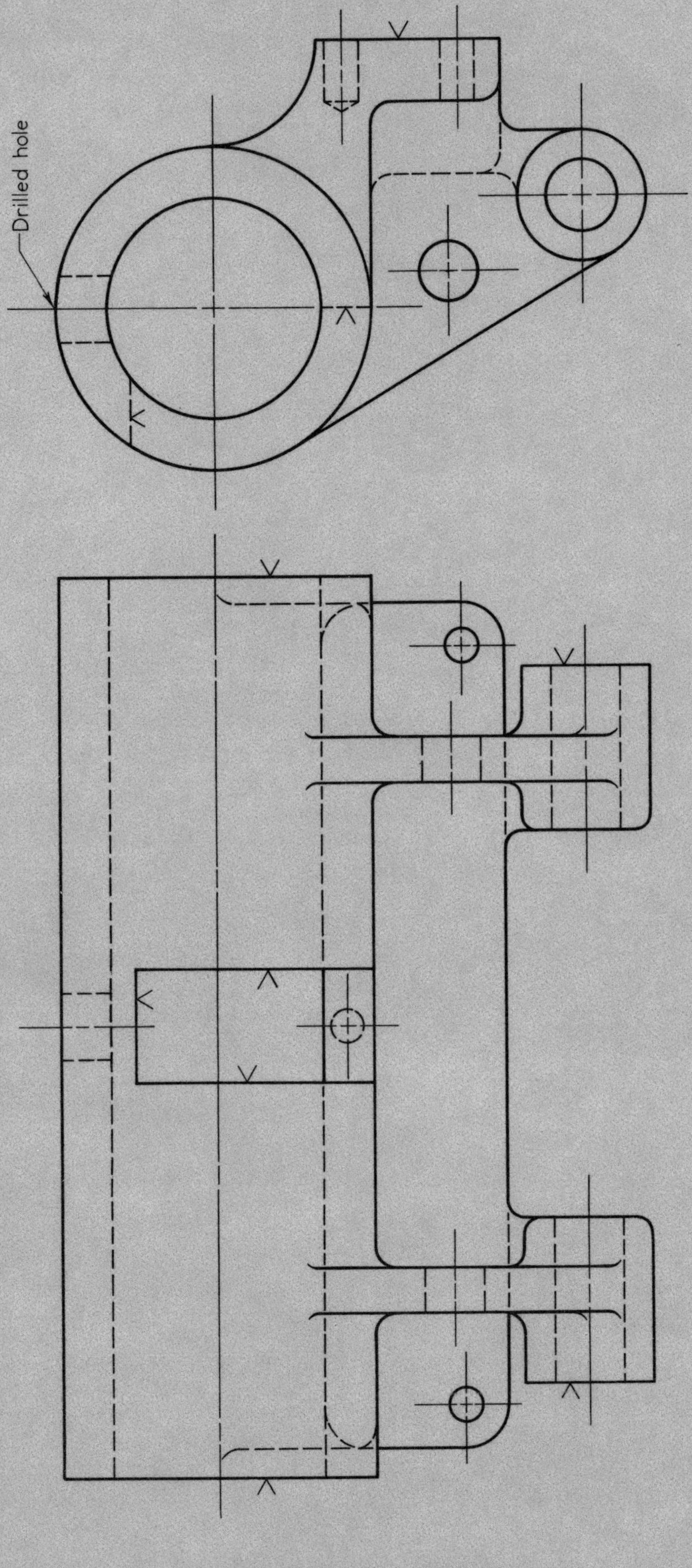
Drilled hole

CONTROL BRACKET

Draw front view.

BOTTOM OF DRAWING

SECTION	DRAWN BY:		SHEET 40

CONTROL BRACKET

Draw front view.

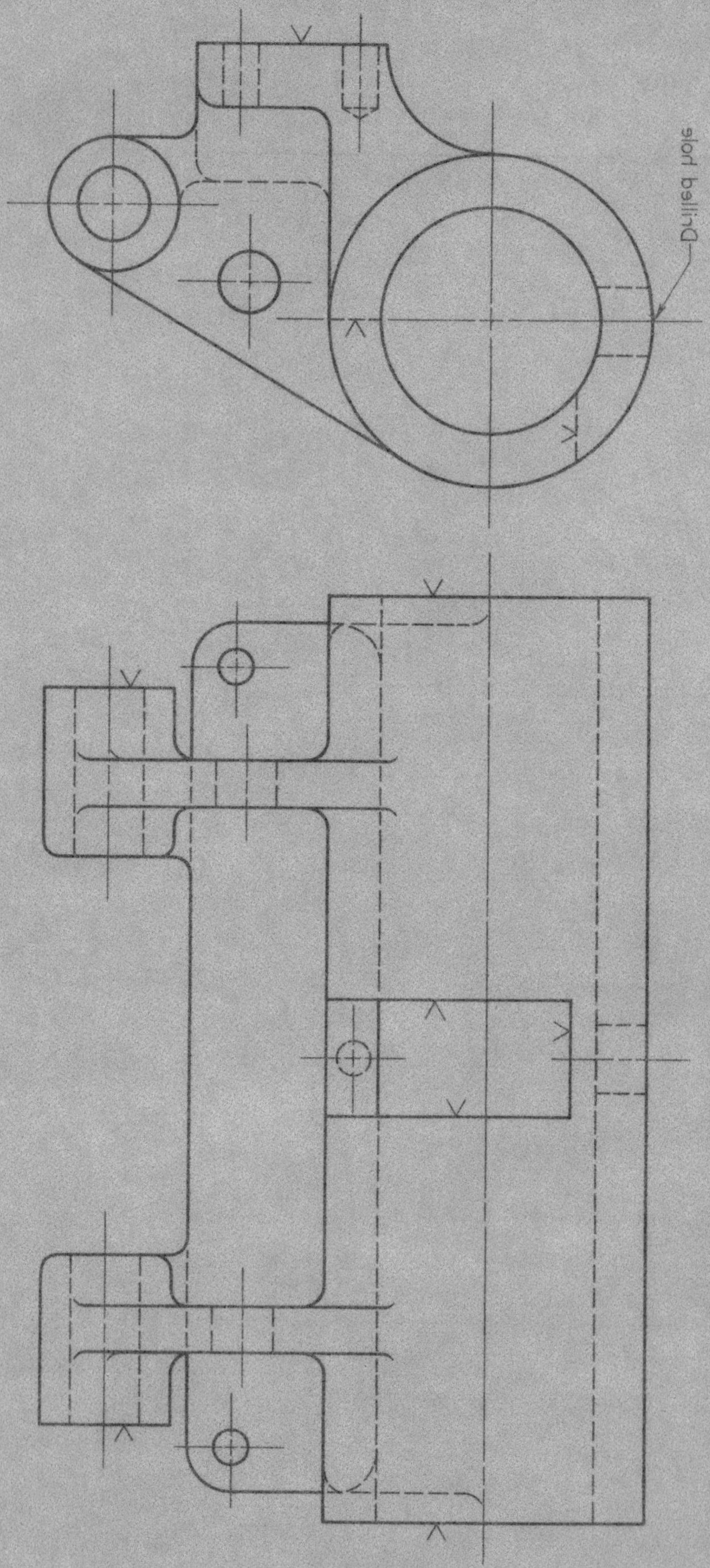
Drilled hole

BOTTOM OF DRAWING

SECTION	DRAWN BY:		SHEET 40

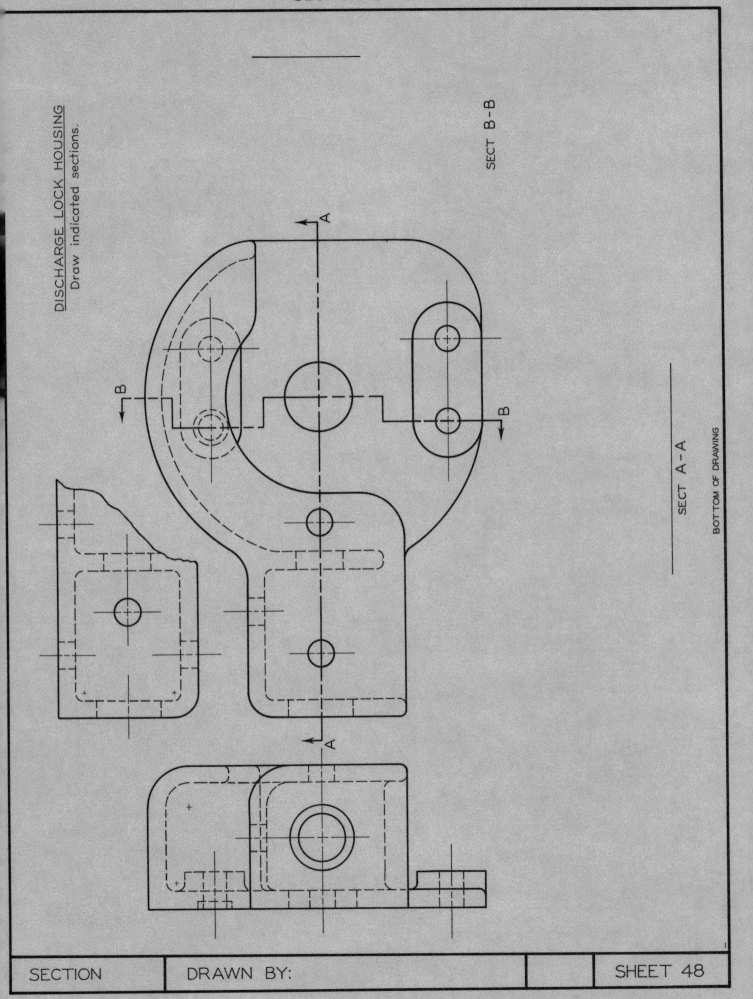

DISCHARGE LOCK HOUSING

Draw indicated sections.

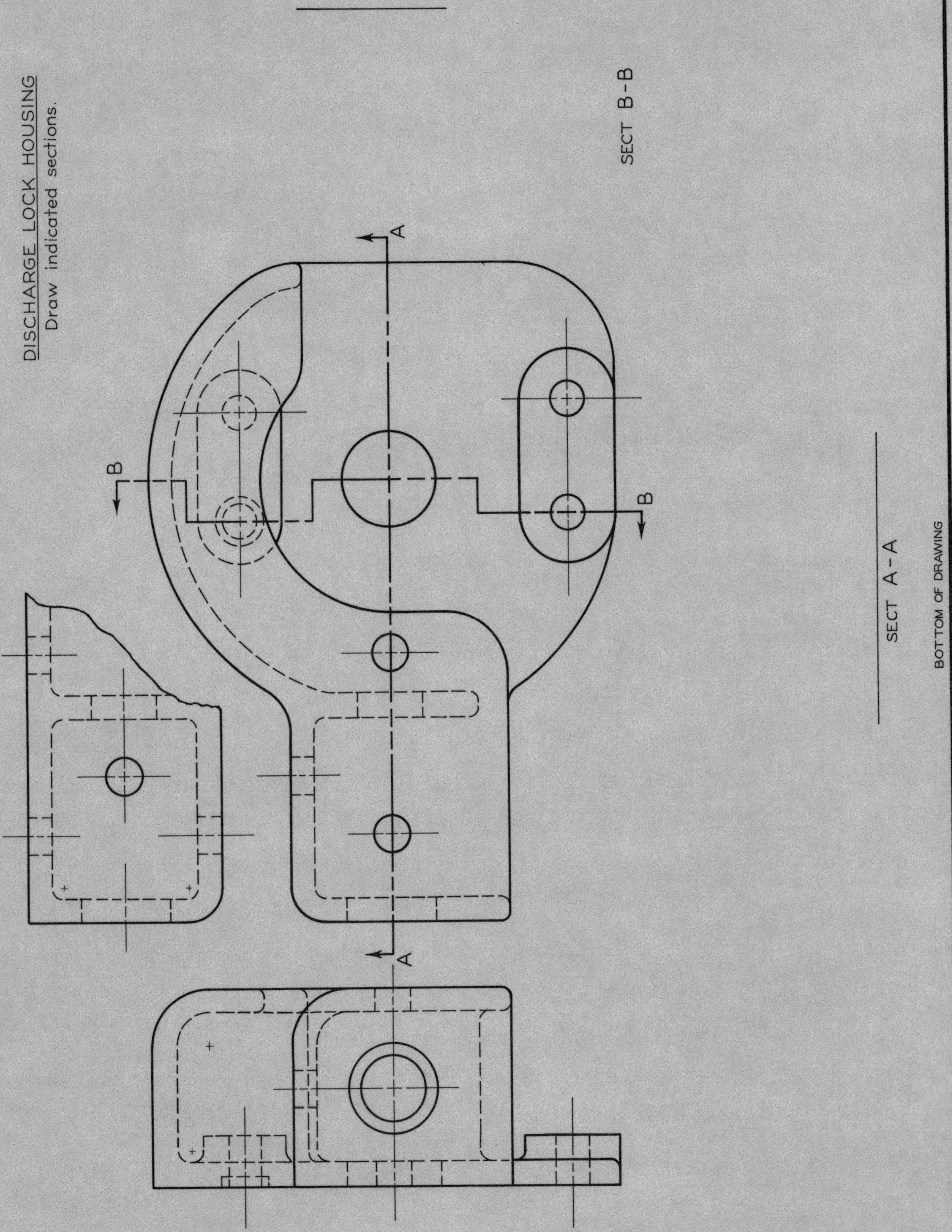
SECT B-B
A
B
B
A
SECT A-A
BOTTOM OF DRAWING

SECTION	DRAWN BY:		SHEET 48

DISCHARGE LOCK HOUSING

Draw indicated sections.

SECT B-B

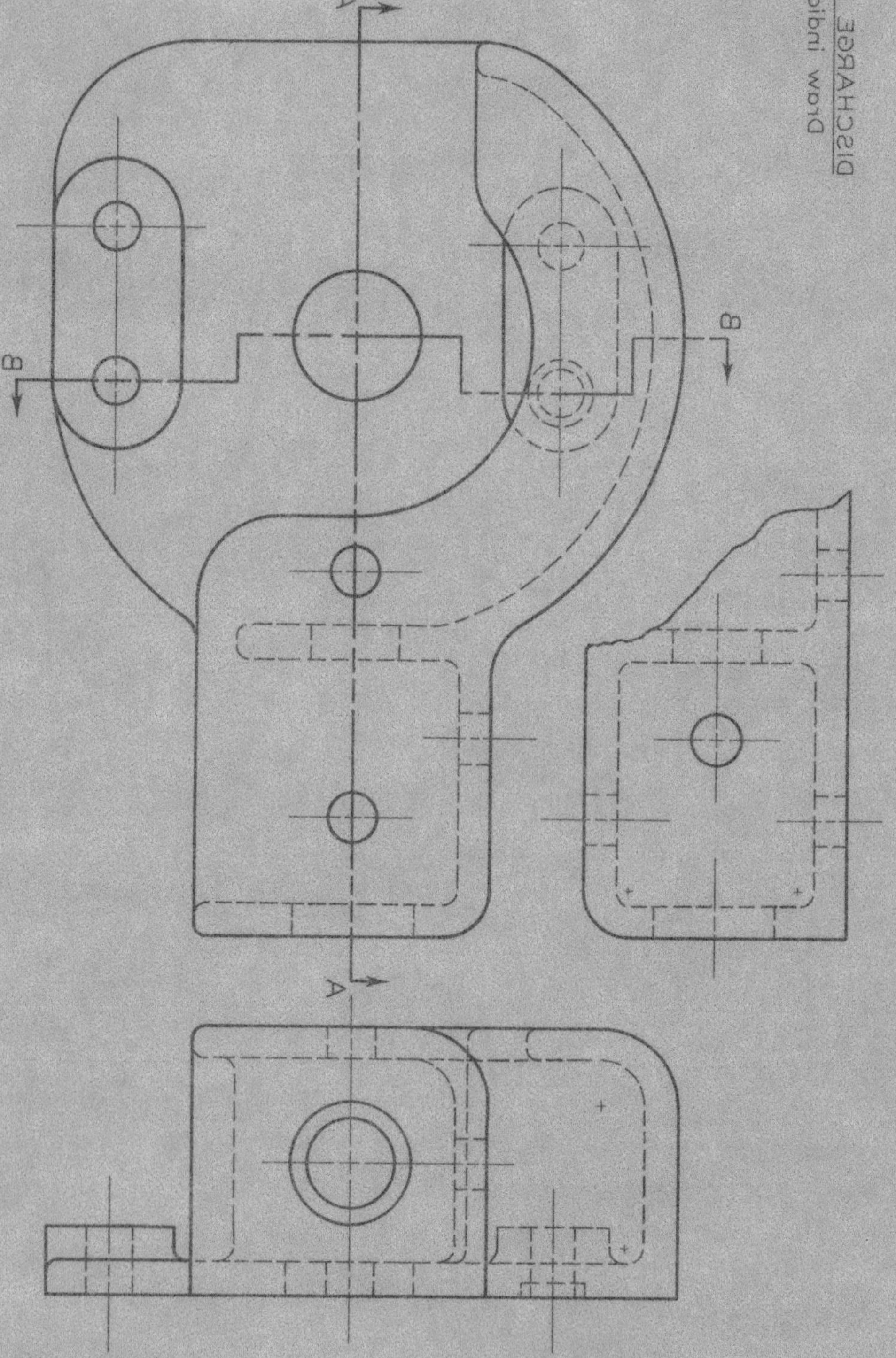
A
A
B
B

SECT A-A

BOTTOM OF DRAWING

SECTION	DRAWN BY:		SHEET 48

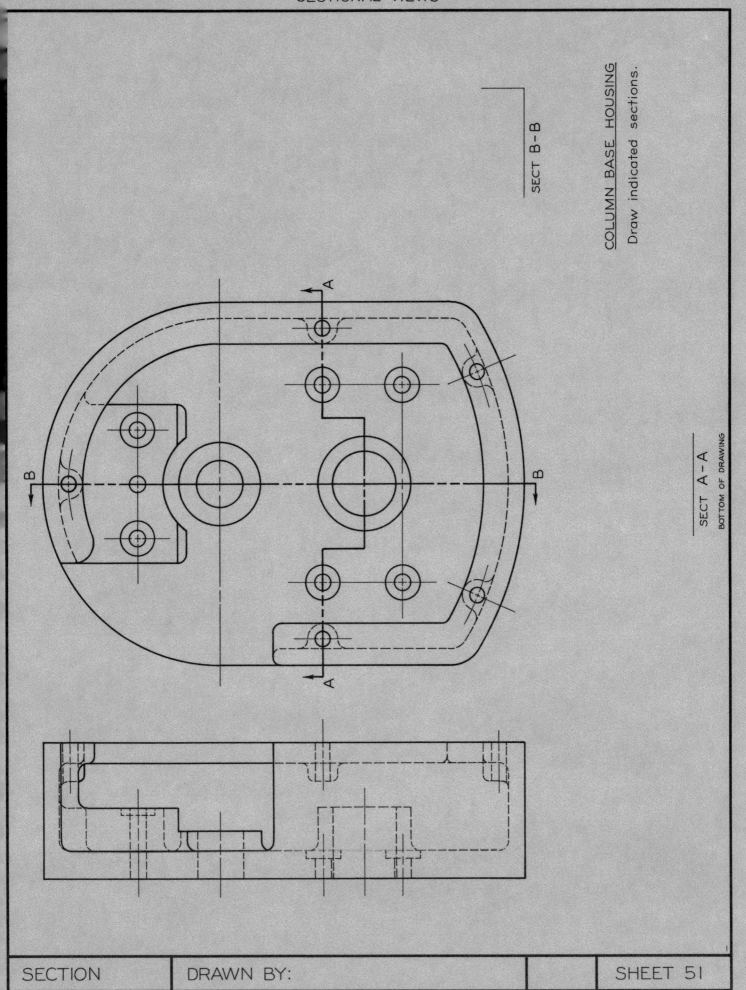

COLUMN BASE HOUSING

Draw indicated sections.

SECT B-B

SECT A-A

BOTTOM OF DRAWING

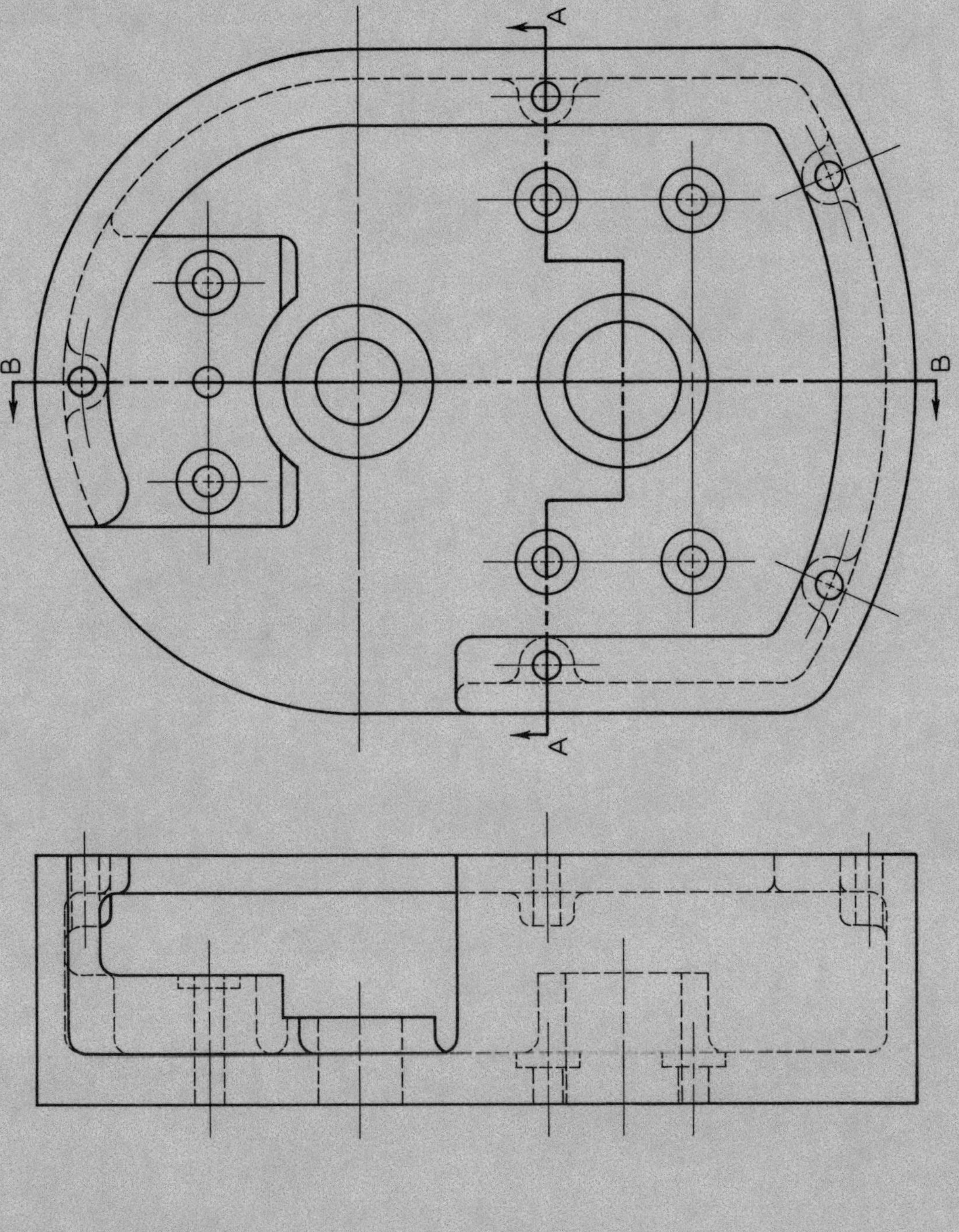
A
A
B
B

SECTION	DRAWN BY:		SHEET 51

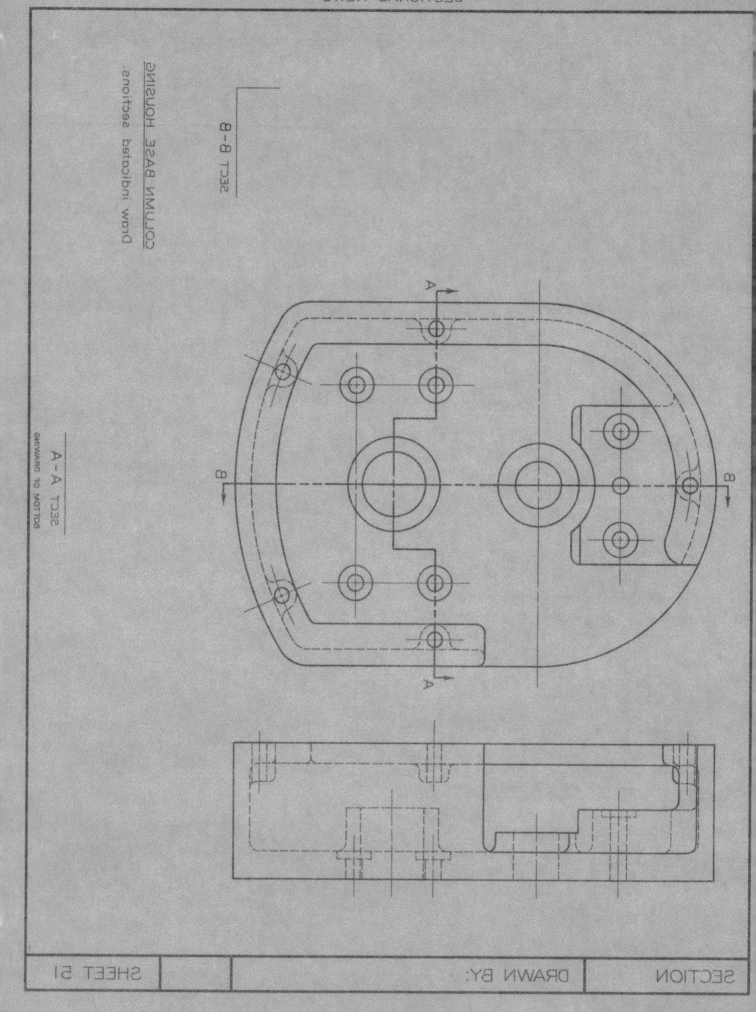

COLUMN BASE HOUSING

Draw indicated sections.

SECT B-B

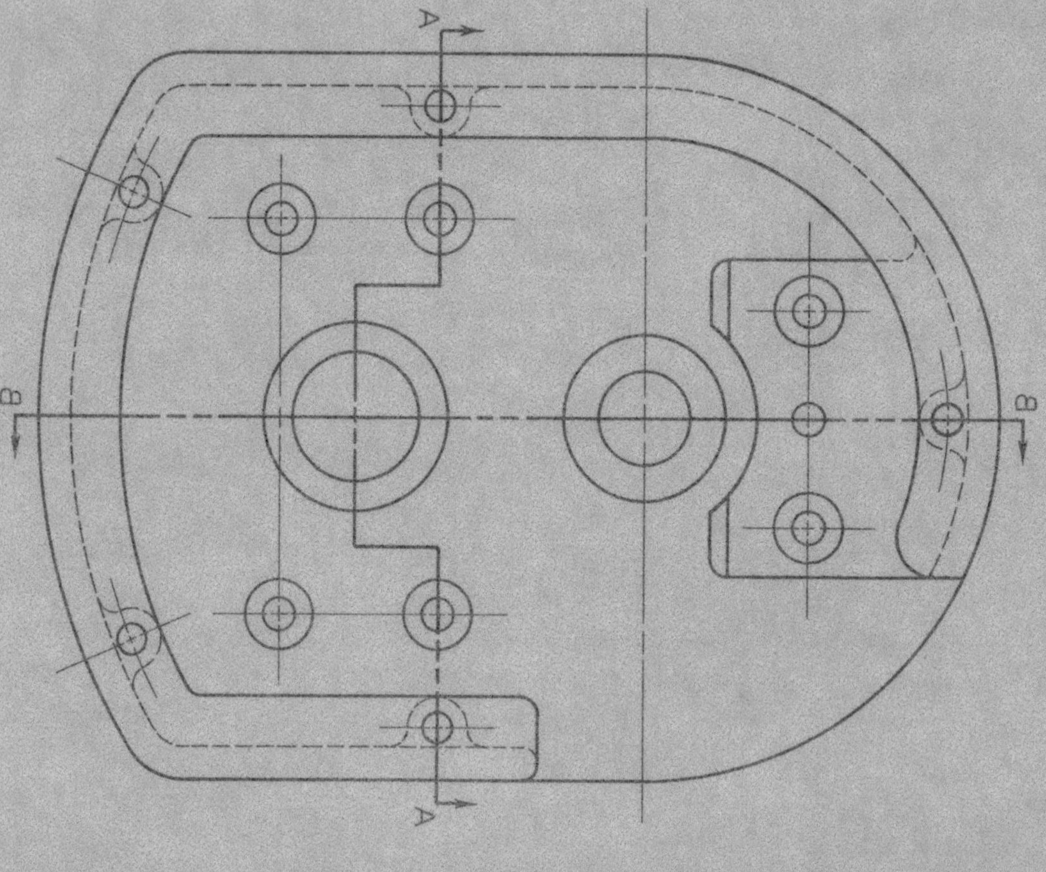
A
A
B
B

SECT A-A

BOTTOM OF DRAWING

SECTION	DRAWN BY:		SHEET 51

1

ROD GUIDE

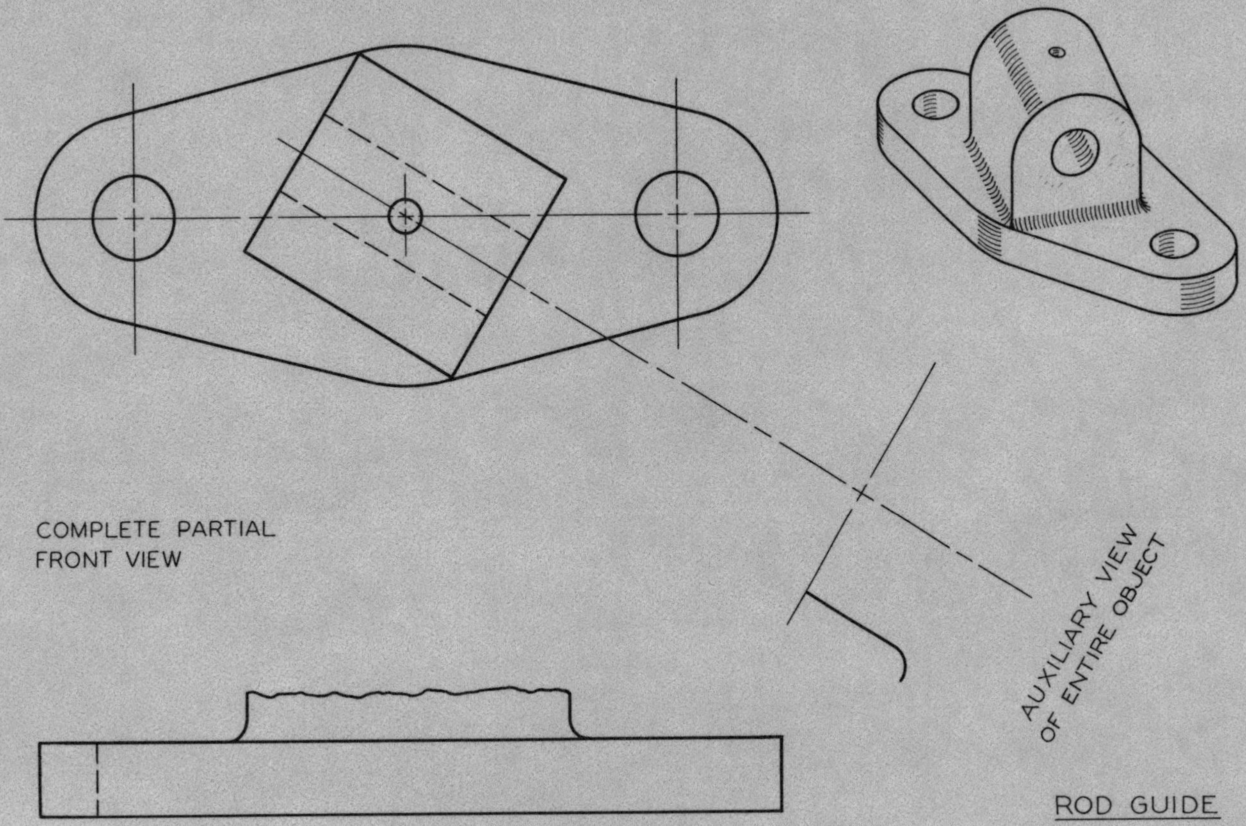

COMPLETE PARTIAL FRONT VIEW

AUXILIARY VIEW OF ENTIRE OBJECT

2

BRUSH HOLDER

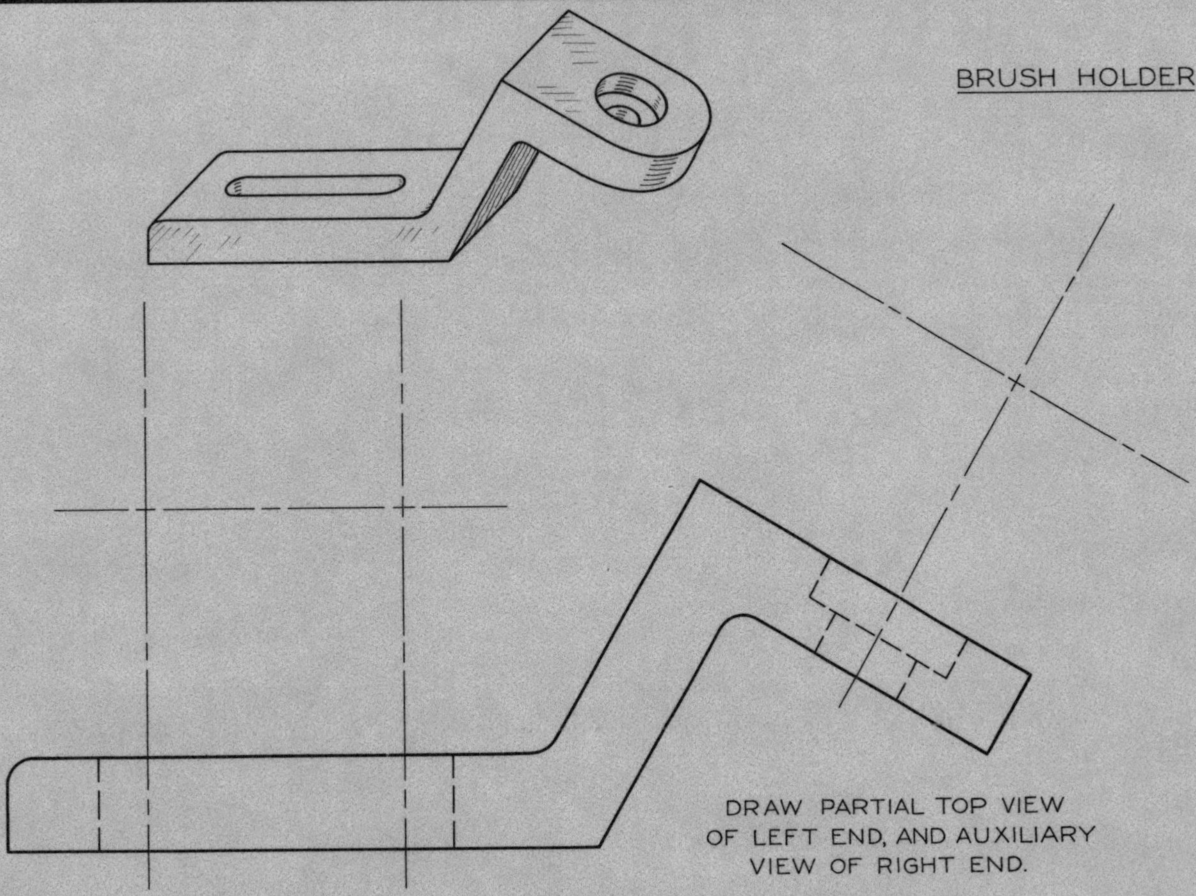

DRAW PARTIAL TOP VIEW OF LEFT END, AND AUXILIARY VIEW OF RIGHT END.

SECTION	DRAWN BY:		SHEET 53

(1)

ROD GUIDE

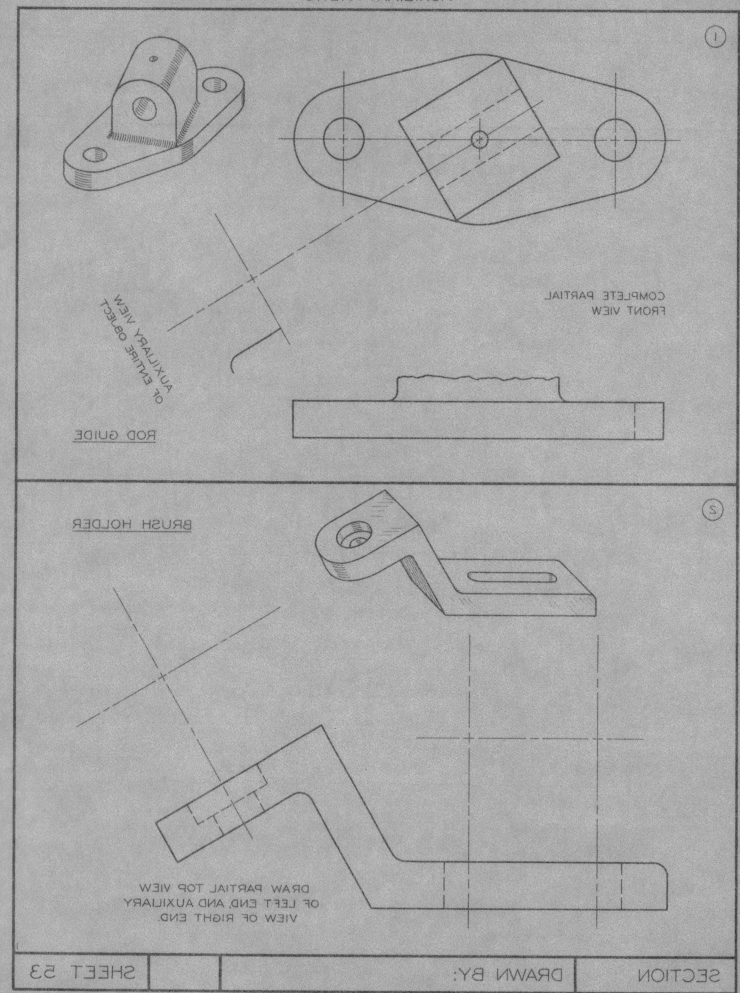

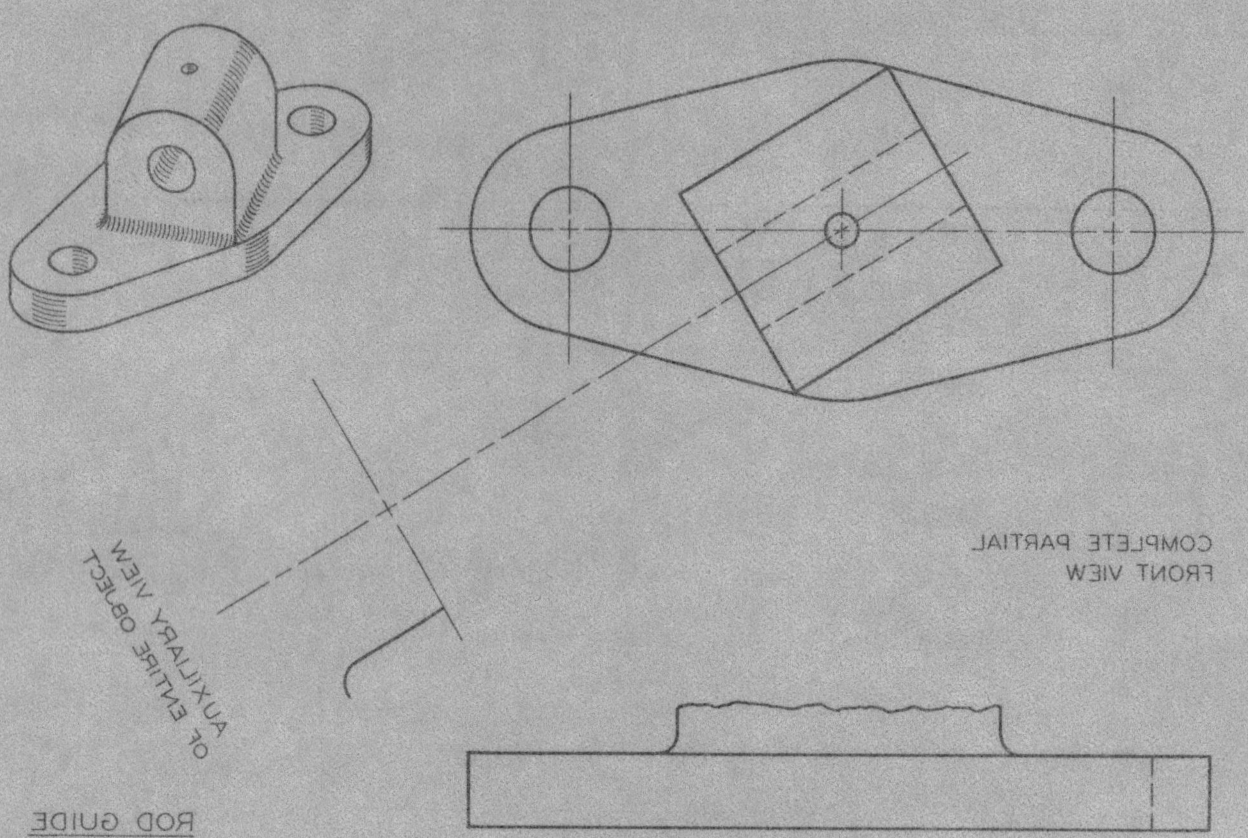
COMPLETE PARTIAL
FRONT VIEW
AUXILIARY VIEW
OF ENTIRE OBJECT

(2)

BRUSH HOLDER

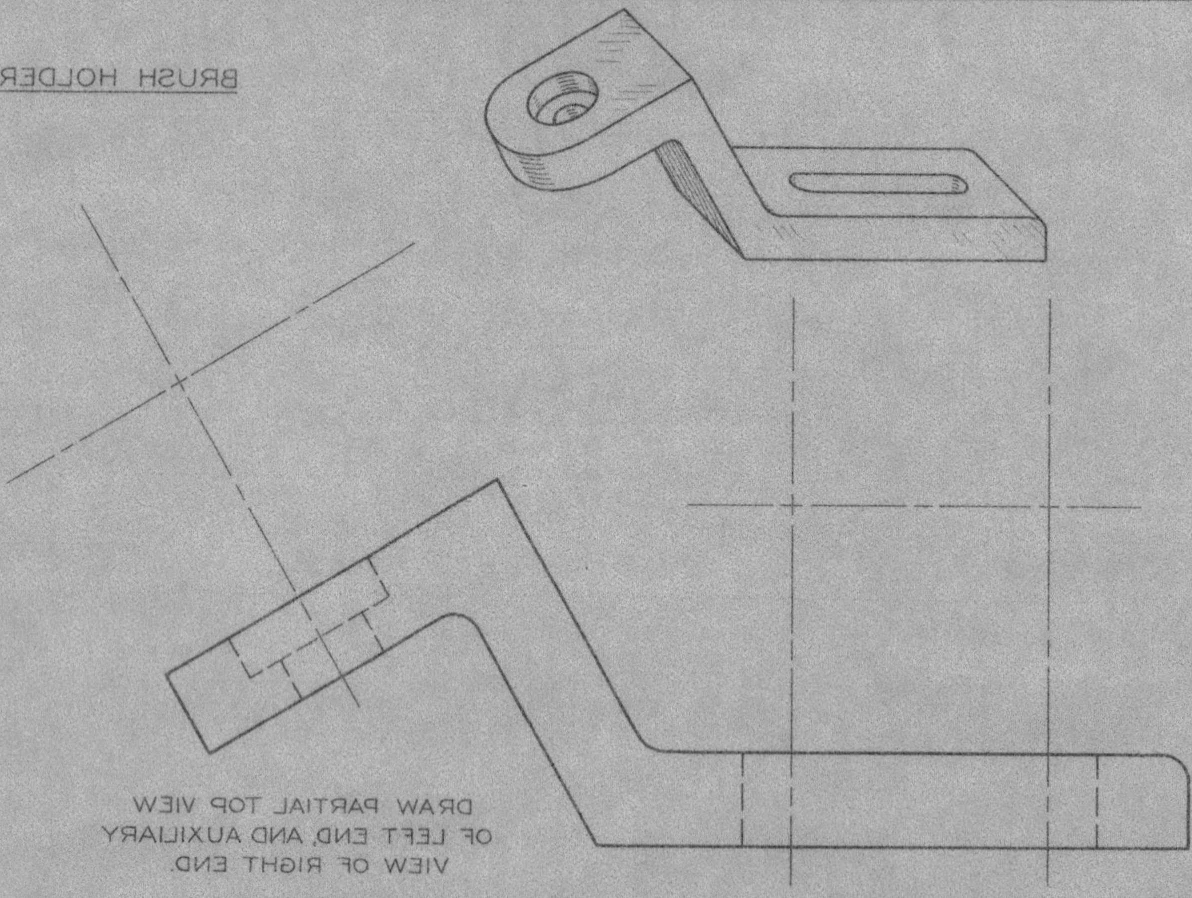

DRAW PARTIAL TOP VIEW
OF LEFT END, AND AUXILIARY
VIEW OF RIGHT END.

SECTION	DRAWN BY:		SHEET 53

(1)

ALIGNMENT CAM

Draw complete auxiliary view showing true size of surface A. Draw right-side view.

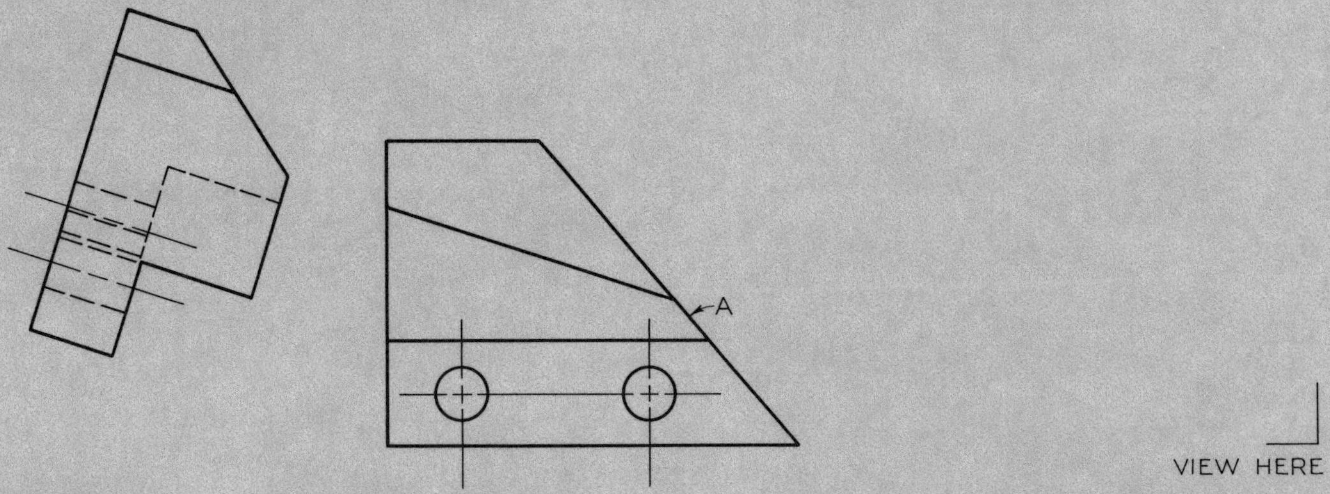
A
VIEW HERE

(2)

SLOTTED BASE

Draw auxiliary section.

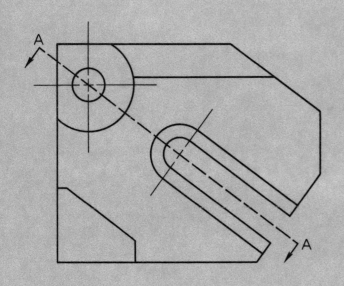

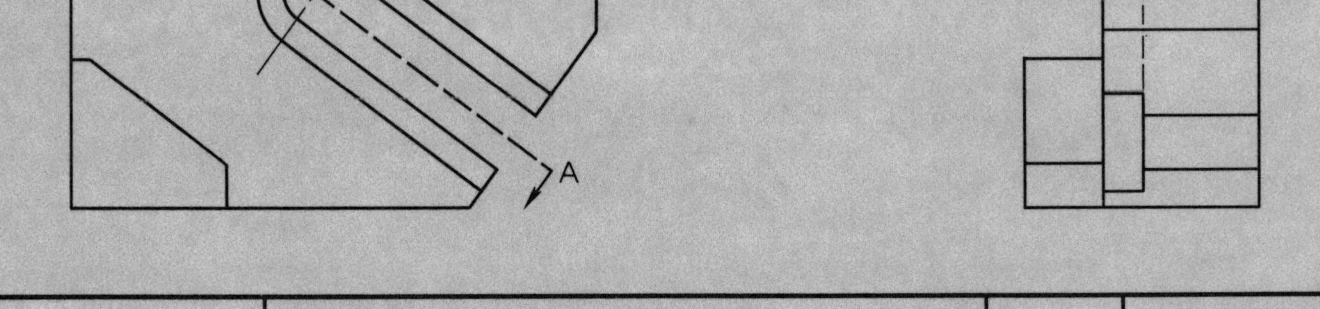
A
A

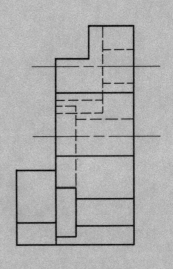

SECTION	DRAWN BY:		SHEET 56

1

ALIGNMENT CAM

Draw complete auxiliary view showing true size of surface A. Draw right-side view.

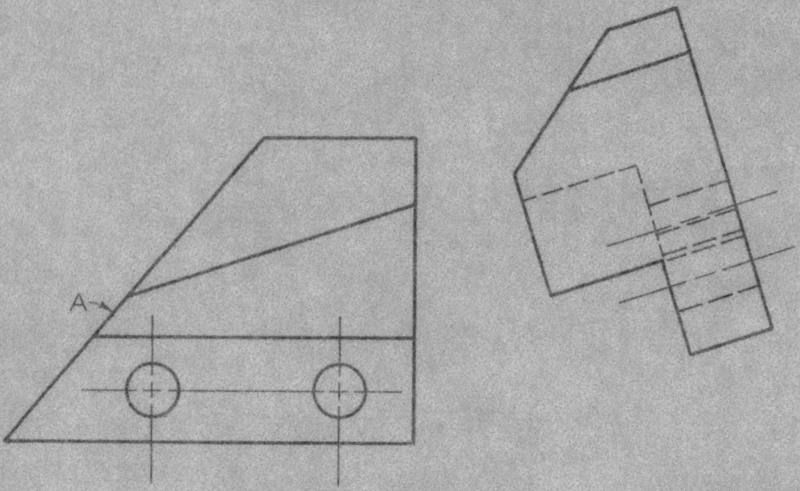
A

VIEW HERE

2

SLOTTED BASE

Draw auxiliary section.

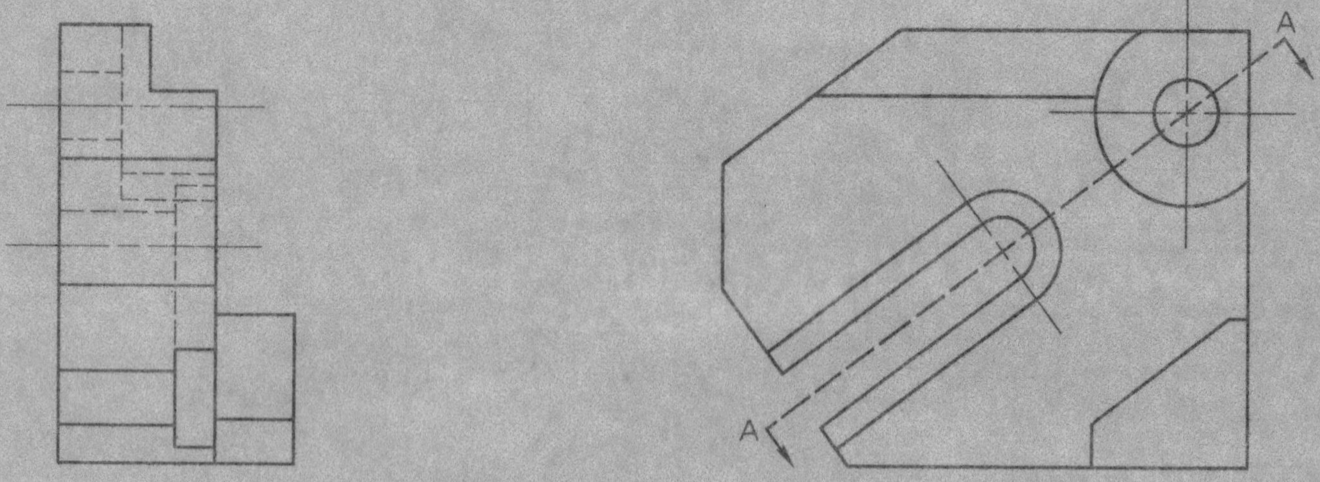
A
A

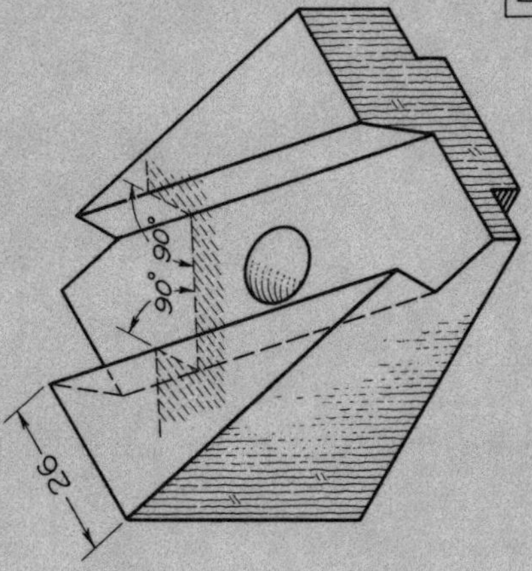
26
90°
90°

METRIC

SLOTTED GUIDE

Draw complete primary auxiliary view, followed by complete secondary auxiliary view showing end of diagonal slot, and complete the right-side view.

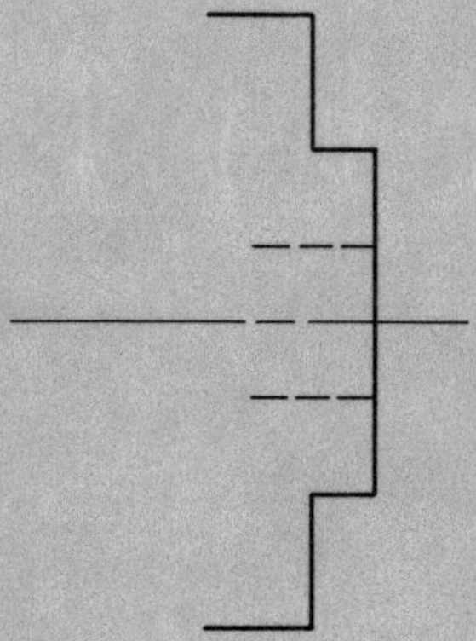

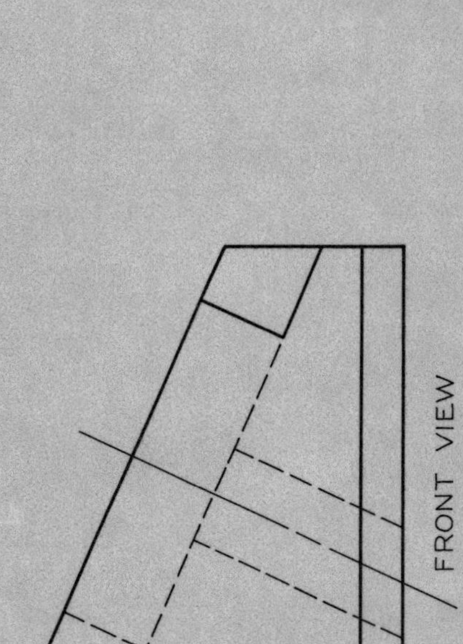

FRONT VIEW

BOTTOM OF DRAWING

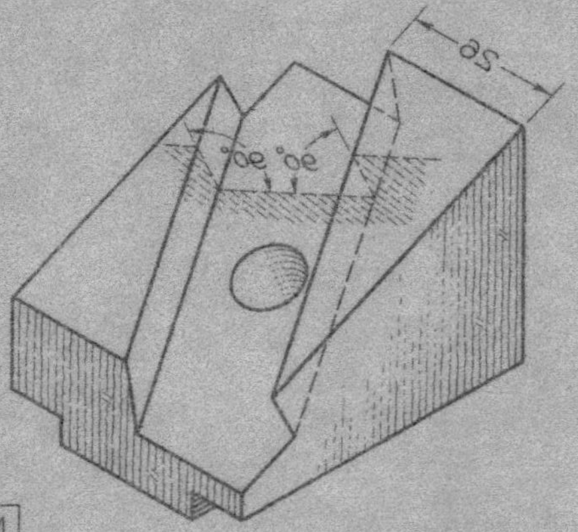
26
90°
90°

METRIC

SLOTTED GUIDE

Draw complete primary auxiliary view, followed by complete secondary auxiliary view showing end of diagonal slot, and complete the right-side view.

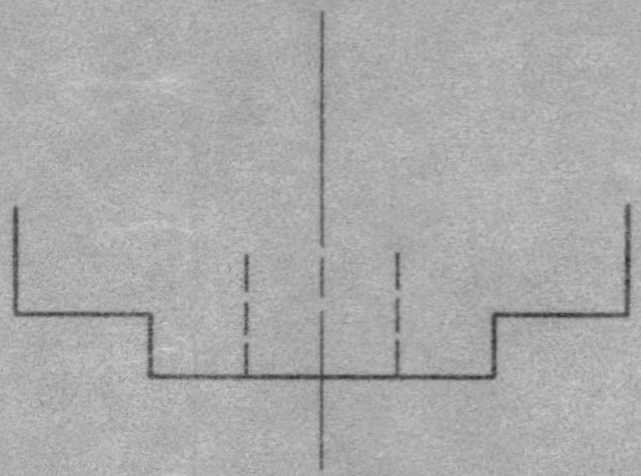

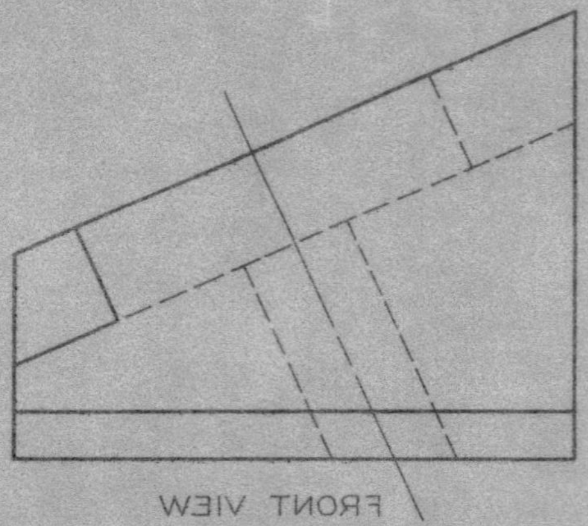

FRONT VIEW

BOTTOM OF DRAWING

SECTION	DRAWN BY:		SHEET 61

(1)

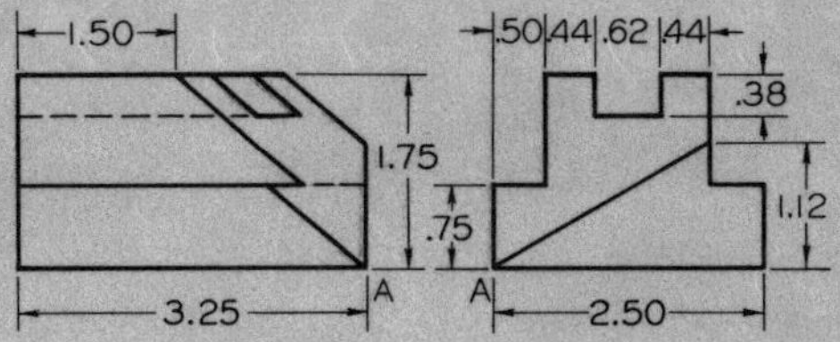
1.50
.50
.44
.62
.44
.38
1.75
.75
1.12
A
A
3.25
2.50

ADAPTER

Draw isometric drawing.

A

(2)

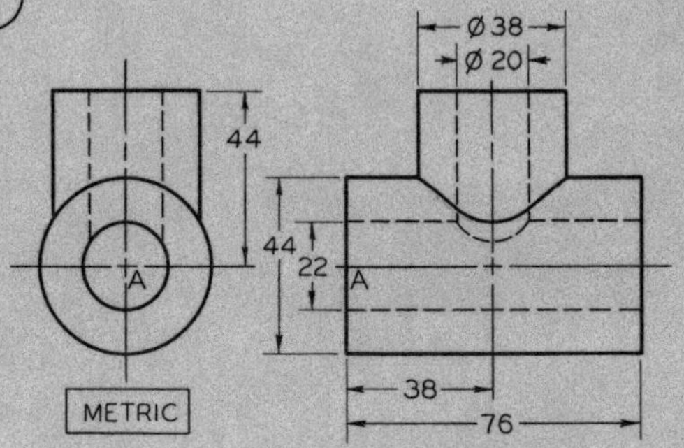
Ø 38
Ø 20
44
44
22
A
A
38
76
METRIC

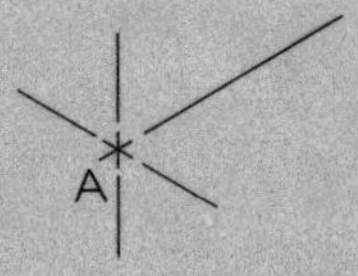
A

TEE

Draw isometric drawing.

1

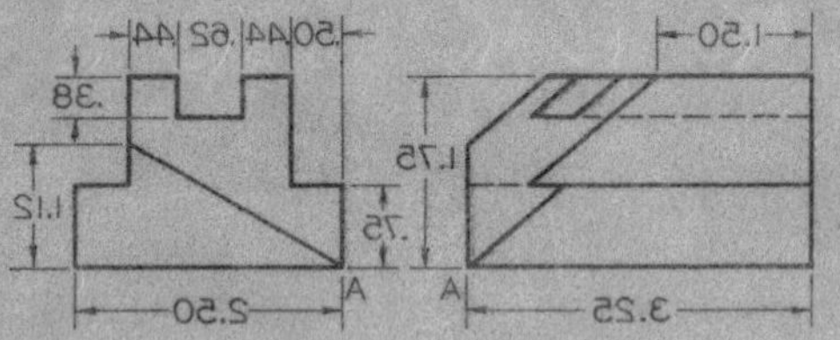
.50
.44
.62
.44
.38
1.12
.75
1.75
1.50
2.50
3.25
A
A

ADAPTER

Draw isometric drawing.

A

2

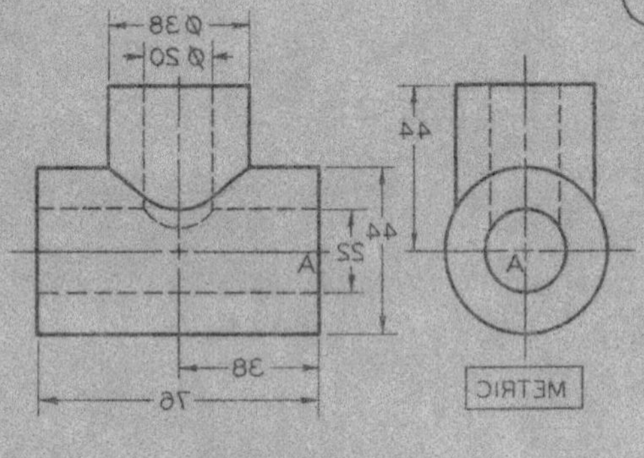
Ø38
Ø20
44
44
22
38
76
A
A
METRIC

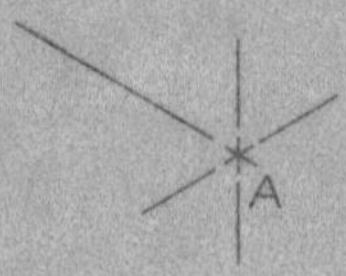
A

TEE

Draw isometric drawing.

SECTION	DRAWN BY:			SHEET 71

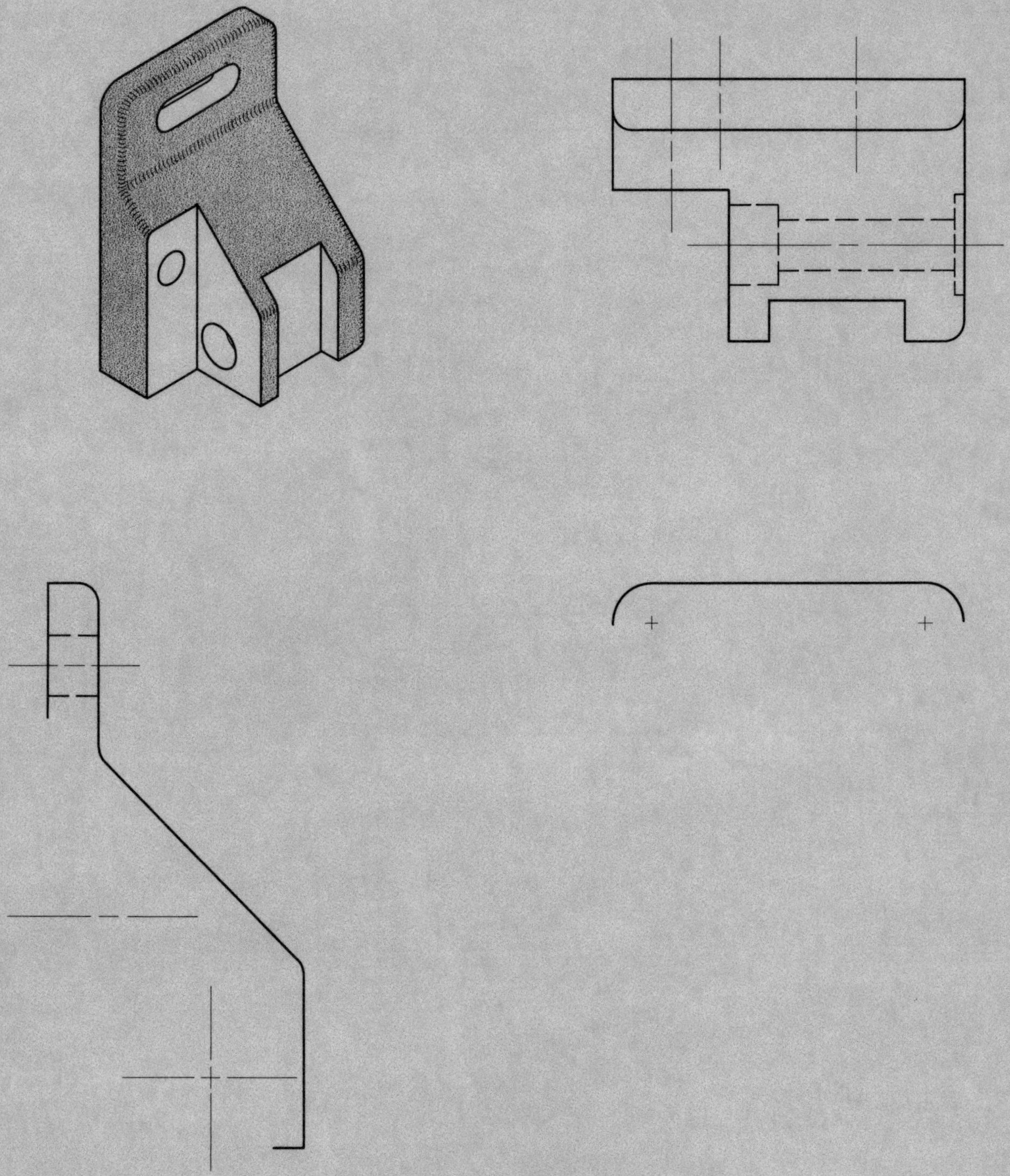

PART NAME – – USED ON – –			THE CONSOLIDATED AUTOMATIC MACHINE CORPORATION MILWAUKEE, WIS.			
DRAWN BY – –	FILE NO. – –	DATE – –	MATERIAL – –	NO. REQD – –	SCALE – –	DRAWING SHEET 80

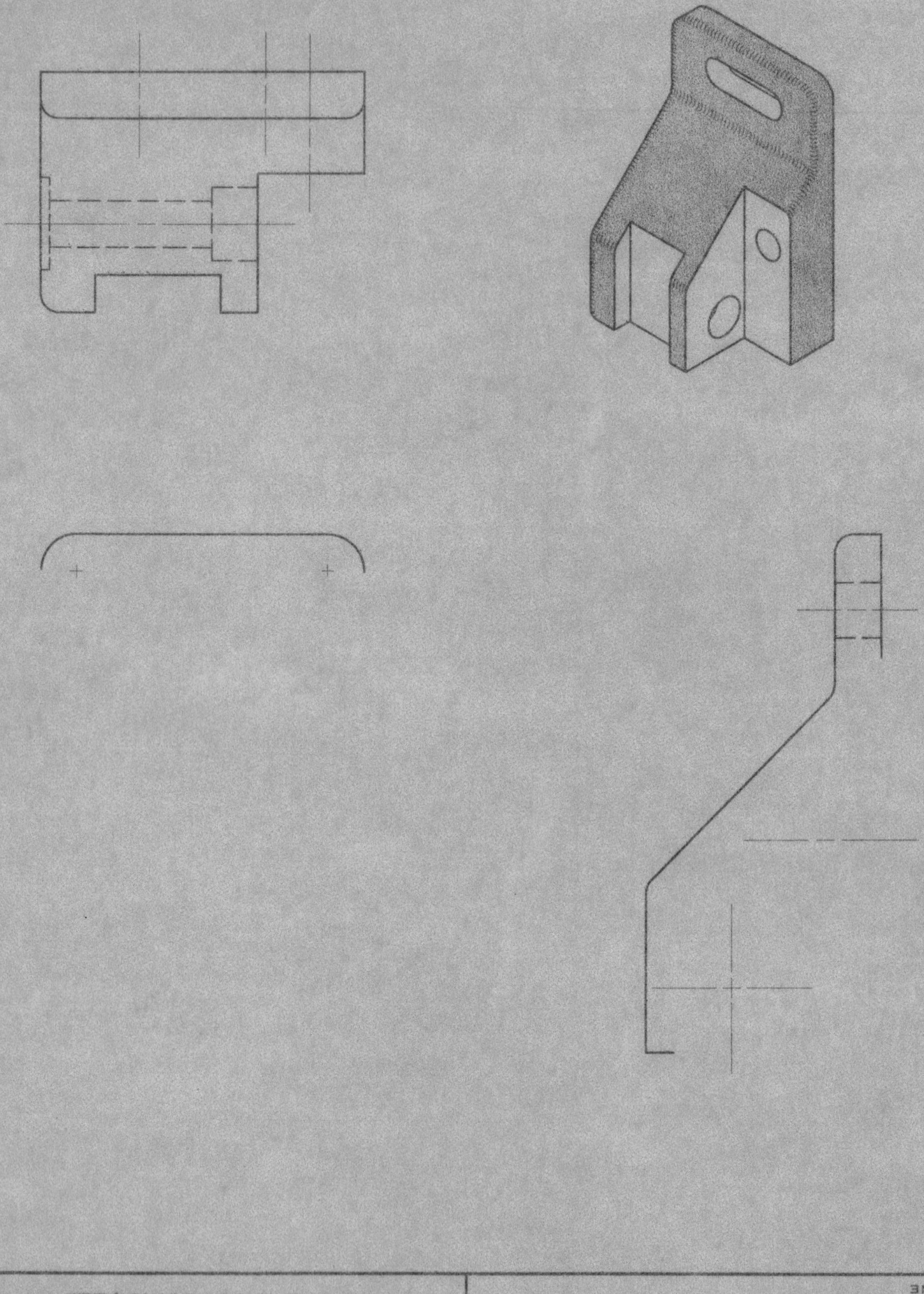

THE CONSOLIDATED
AUTOMATIC MACHINE CORPORATION
MILWAUKEE, WIS.

PART NAME
USED ON

DRAWING	SCALE	NO. REQD	MATERIAL	DATE	FILE NO.	DRAWN BY
SHEET 80						

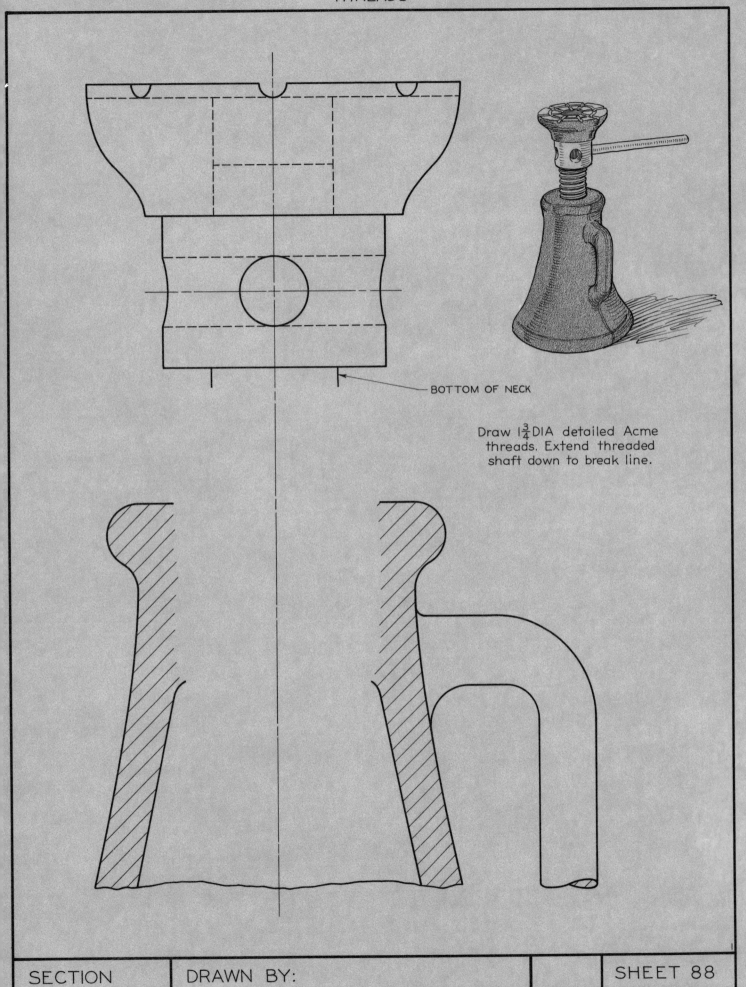

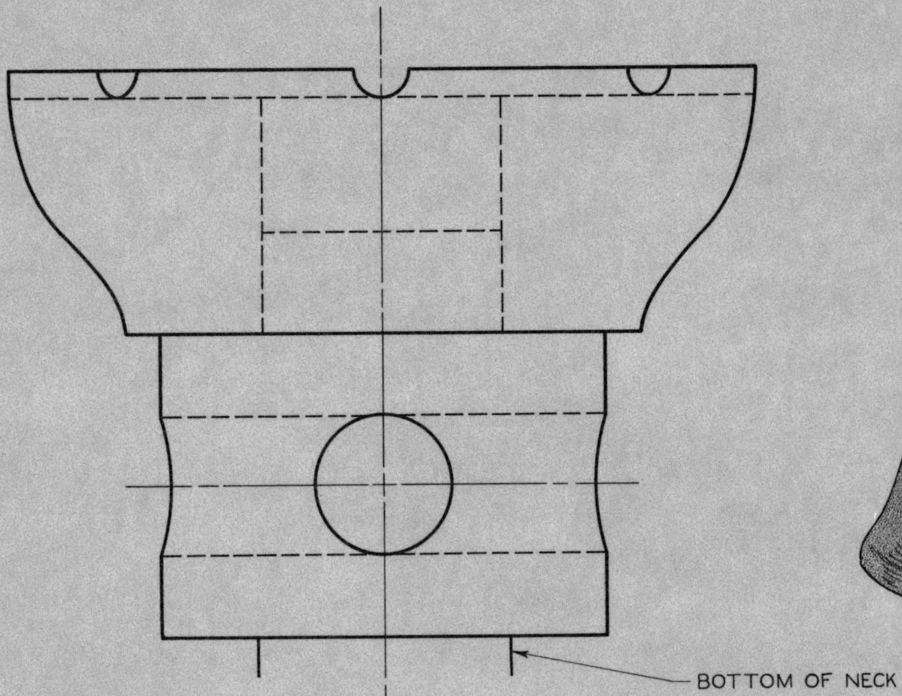
BOTTOM OF NECK

Draw $1\frac{3}{4}$ DIA detailed Acme threads. Extend threaded shaft down to break line.

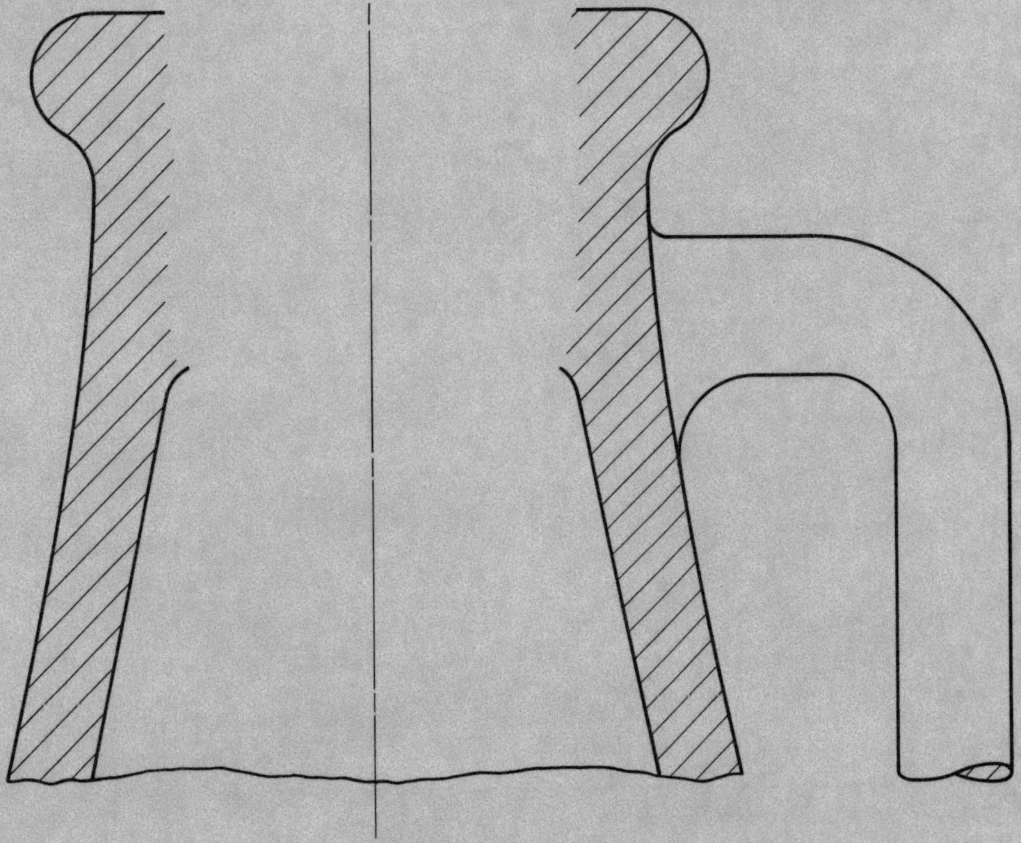

SECTION	DRAWN BY:		SHEET 88

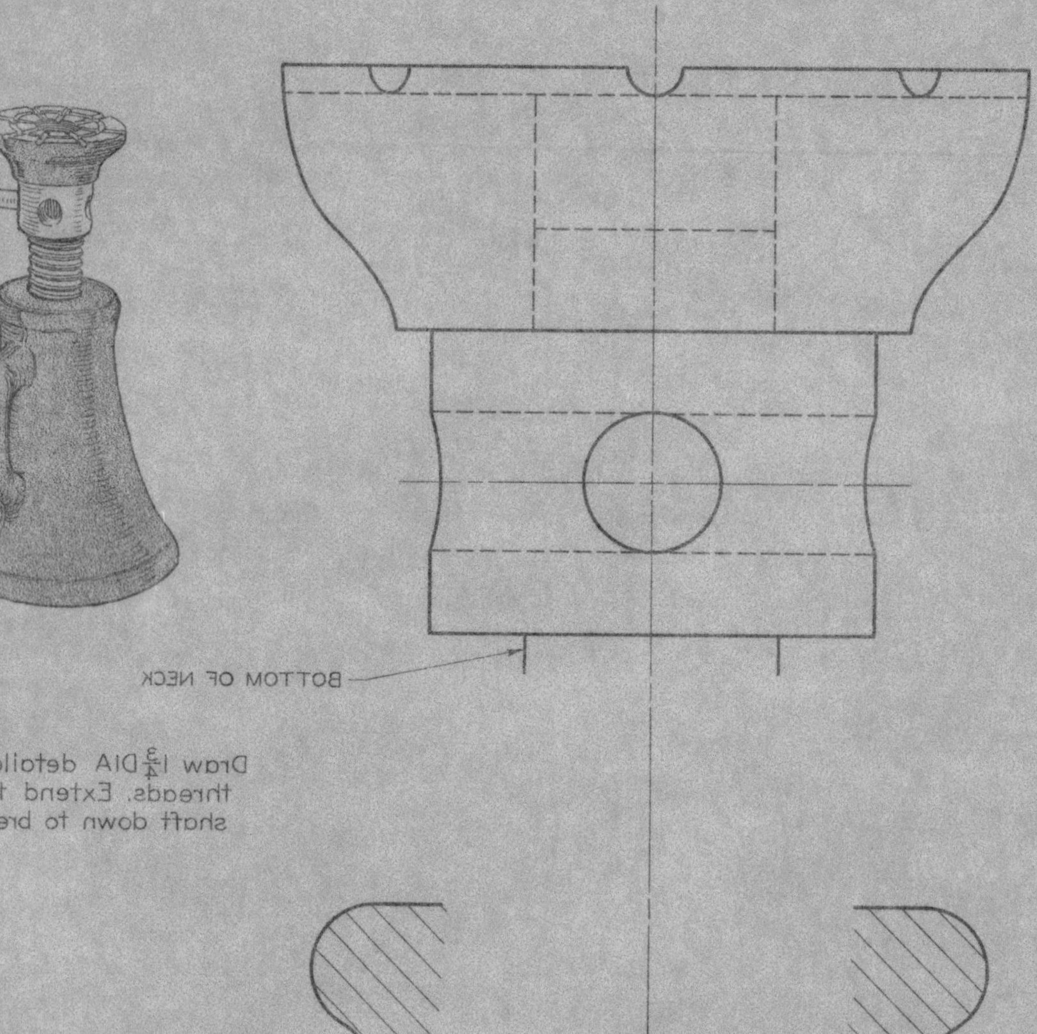
BOTTOM OF NECK

Draw $1\frac{3}{4}$ DIA detailed Acme threads. Extend threaded shaft down to break line.

SECTION	DRAWN BY:		SHEET 88

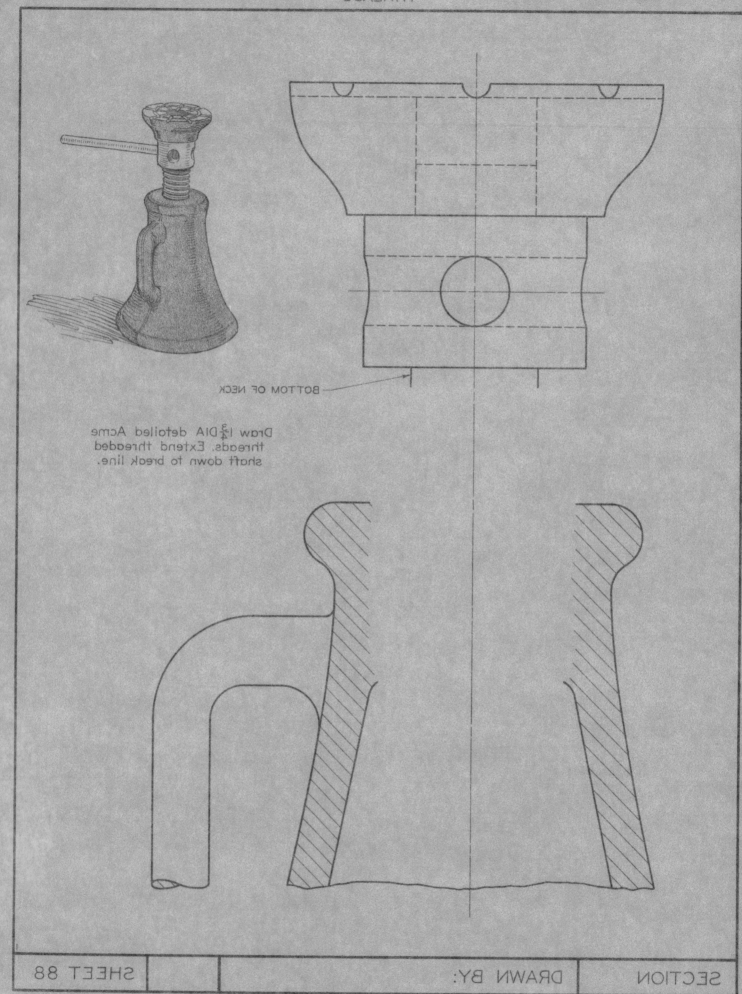

SECTION	DRAWN BY:		SHEET

SECTION	DRAWN BY:			SHEET

SECTION	DRAWN BY:		SHEET

SECTION	DRAWN BY:			SHEET